Above: An aerial view of the supercell thunderstorm that produced the La Plata tornado, April 28, 2002. The bulging dome of clouds extending above the supercell's flat, anvil top is caused by a very intense updraft.This photograph was taken 15-30 minutes after the photograph on the previous page. *Steven Maciejewski*

Previous page: A large tornado spins on the Chesapeake Bay southeast of Long Beach, St. Leonard, Maryland, April 28, 2002. The tornado was associated with the supercell thunderstorm (above) that devastated La Plata, Maryland. *Ted L. Dutcher*

Next page: Baseball-sized hail that fell near Hughesville, Maryland, April 28, 2002. The hail fell from the supercell thunderstorm (above) that produced the tornado in La Plata, Maryland. *NOAA Photo Contributor*

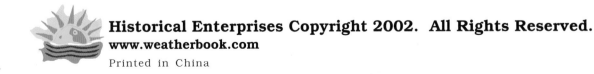

WASHINGTON WEATHER
The Weather Sourcebook for the D.C. Area

TK,
Any time you need
a forecast -- revenue or
weather -- let me
know !

Kevin   Dan   Andy
Ambrose   Henry   Weiss

DEDICATION

Kevin:
To my wife, Elisa, and sons, Bradley and Michael, whose support helped make this effort possible.

Dan:
To Kerry, Logan, and Brooke for their love, inspiration, and incredible patience during the countless late nights at the computer.

Andy:
To the memory of Louis Allen, the amiable Channel 7 meteorologist of the 1950's and 1960's who got a whole generation of Washingtonians hooked on weather.

A newspaper photographer is about to be pelted by snowballs at 13th and L Streets, NW, March 16, 1950. The brief snowstorm dropped 1.8 inches of snow on Washington, with a high temperature of 48°F and a low temperature of 28°F. *Copyright Washington Post; Reprinted by permission of the D.C. Public Library*

Author's Note: Of the thousands of photographs that were reviewed for *Washington Weather*, the above photo is one of my favorites. It captures a moment of childhood joy that even a small snowfall can create. The snow melted just hours after the photograph was taken, but this scene, captured on film over 50 years ago, remains frozen in time. – KA

TABLE OF CONTENTS

Preface 7

Introduction 9

Chapter 1: Early Weather 13

Chapter 2: Winter Storms 43

Chapter 3: Cold Waves 111

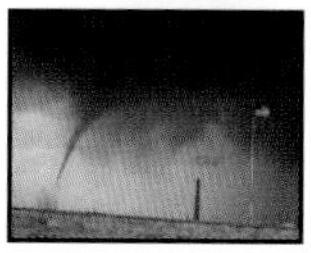

Chapter 4: Severe Weather 131

Chapter 5: Tropical Weather 175

Chapter 6: Floods 197

Chapter 7: Heat Waves 225

Appendix 245
Bibliography 247
Index 249
Acknowledgements 251

A thunderstorm near Alexandria, Virginia sets a stormy backdrop for the Washington Monument, August 2, 2002. The nearly stationary storm caused flash flooding in Alexandria and southeast Fairfax County. This photograph was taken at dusk from Constitution Avenue. (See page 164 for a photograph sequence of the storm.) *Kevin Ambrose*

PREFACE

Having spent most of my adult life in the Washington area, I can say, without hesitation, that this is one of the most difficult parts of the country to forecast weather. Of course, the same can be said of Richmond, Baltimore, Philadelphia, New York and Boston. You see, Washington and the other cities share some common natural boundaries — the mountains to the west and the Atlantic Ocean to the east. Throw in the Chesapeake Bay and you have created our own, unique climatic zone.

And what a climate we have! There are four distinct seasons to enjoy here, each with its own trademark unpredictability. Our occasional summer heat waves are as oppressive as any you will find along the East Coast. Blizzards, when we have them (six in my memory), are as deep and wind-swept as any you will find at our latitude. Autumn is my favorite season. October is our standout weather month. It is usually a pleasant extension of late summer — bonus days before the first freeze. However, every now and then autumn features direct effects of hurricanes and tropical storms. And spring, sometimes it arrives early and throws the Cherry Blossoms into blooming confusion. Some years it is late March, other years it is early to mid-April when these pink national treasures come into full bloom around the Tidal Basin. One wicked spring thunderstorm, though, can strip the blossoms from the branches. And sometimes those storms spawn tornadoes. From F0 to F4, we have seen them, even in our close-in suburbs.

Our climate is never boring. I cannot imagine any other city that offers such a variety of weather patterns. Kevin, Andy and Dan have captured the power and intensity of extreme Washington weather. The stories, explanations, history, photographs and graphics are first class. So, whether you are a weather professional or hobbyist, or just someone who is fascinated by our predictably unpredictable weather, you will treasure this work. And you might even discover that your parents and grandparents were not exaggerating when they recounted some of the historic storms to hit this area. Maybe they really did walk to school in waist-deep snow or were without electricity for two weeks.

Doug Hill
Chief Meteorologist
WJLA
Washington, D.C.

INTRODUCTION

The Mid-Atlantic has dealt with some incredible storms over the last couple centuries from great floods to hurricanes, heat waves to blizzards, and hailstorms to tornadoes. The Washington area sees it all. While you need to dig back in time to find when ice formed on the Chesapeake Bay thick enough for people to walk from Annapolis to Kent Island, you don't have to go back so far to find many of the other types of significant weather that has impacted the area.

The mid-1990s brought a rough string of storms. In June of 1995, a thunderstorm complex dropped over 20 inches of rain on Madison County, Virginia. Parts of the mountainsides gave way and came down in debris flows washing away homes, roads and bridges. Amazingly only 3 people were killed. Over 80 people had to be rescued — plucked off rooftops by helicopters. The flood event was reminiscent of the disaster that struck Nelson County Virginia (south of Charlottesville) in August 1969 when the remnants of Hurricane Camille came through. Two feet of rain fell after dark and by morning, over 150 people were dead. It was one of the country's worst flash floods.

Some areas that experienced flooding in June of 1995 would flood two more time in 1996. Both of these events were also remarkable. January 1996 began with the biggest snowstorm to hit the area in 13 years. Two to three feet of snow blanketed the region, shutting down the federal government for four days and schools for over a week. Then, on January 19, a warm air mass moved in. The warm air over the cold snow pack caused an eerie fog to form that was so dense in some areas that you could not see beyond the nose of your car. By morning, bare ground was showing. All the snow had disappeared in just one night! The rivers and streams couldn't handle the sudden runoff of melted snow. They flooded. People, determined to get to work on the 20th, tried to drive through the cold floodwaters. They had to be rescued and treated for hypothermia.

Just nine months later, as some people were just finishing their renovations from the January flood, along came Hurricane Fran. It was a Category 4 hurricane when it made landfall near Wilmington, North Carolina, but by the time it reached Virginia it had weakened to a tropical storm. The storm dropped over 15 inches of rain at Big Meadows in Shenandoah National Park. So much rain fell on the mountains on both sides of the Shenandoah Valley that it sent the Shenandoah River into record flood. Almost every road in Page County was washed out. The floodwaters took a while to get down to Washington, D.C. They arrived three days later, Monday morning September 8, during high tide on the Potomac. The flood closed nearby roads such as Rock Creek Parkway and George Washington Parkway snaring traffic and making a bad morning commute.

The mid-1990's were also big years for tornadoes in the region. Maryland normally averages only 4 tornadoes per year. However in 1995, the state set a new record with 25 tornadoes documented. 1996 was not far behind with another 22 tornadoes and 1994 saw 21 tornadoes recorded. Still, Maryland had not seen a killer tornado since the one that struck Dorchester County in 1984 and the Washington area had not scene killer tornadoes since 1929. That good streak came to an end on September 24, 2001. A tornado struck the University of Maryland at College Park killing two students and injuring another 55 people. Then, just six months later, on April 28, 2002, another killer tornado struck Charles and Calvert counties in Maryland. It carved through the heart of La Plata destroying the downtown business district. It was on the ground for over 64 miles destroying over 100 homes and damaging nearly a thousand other buildings. Three people were killed and more than 120 injured. The storm also dropped hailstones up to the size of softballs.

This is just a small sampling of the weather extremes that can strike the Washington area. It is a fascinating region in which to study meteorology and help local governments and communities prepare for what nature deals us.

Barbara Watson
Warning Coordination Meteorologist
National Weather Service
Baltimore-Washington Forecast Office

Facing page: Lightning over Potomac Heights, Maryland, summer of 2000. *John Olexa, Jr.*

The Wye Oak, estimated to be 460 years old, was toppled by a severe thunderstorm, June 5, 2002. The massive tree, located in Wye Mills, Maryland, was one of the largest and most famous oak trees in the U.S. It stood over 100 feet tall and had a circumference of 31 feet. The thunderstorm that downed the tree produced winds in excess of 60 mph. *AP/WIDE WORLD PHOTOS*

WASHINGTON WEATHER
The Weather Sourcebook for the D.C. Area

Snow falling on a wartime White House, March 29, 1942. The snowfall in Washington was 11.5 inches, with a high temperature of 35°F and a low temperature of 32°F. *Copyright Washington Post; Reprinted by permission of the D.C. Public Library.*

Benjamin Franklin's famous lightning experiment, documented to have occurred in June, 1752. Electricity fascinated Ben Franklin, and he conducted many experiments with electrical current. *Library of Congress.*

CHAPTER ONE

EARLY WEATHER
OF THE MIDDLE ATLANTIC
1585-1865

The first documented weather observations for the Middle Atlantic came from early European explorers as they charted the North American coastline and established colonies in eastern North Carolina and southeast Virginia. These navigators described the weather of the Middle Atlantic as being different from that of their homeland of England. They found the weather to be colder in the winter and much hotter in the summer. Also, storms were noted to be much more intense. Captain John Smith wrote the following after settling in Jamestown: "The winds here are variable, but the thunder and lightning to purify the air, I have seldom either seen or heard in Europe."

It was hard for the colonists to explain why the winters in Virginia were colder than that of England, especially since the colonies were located at a latitude well south of England's latitude. The colonists did not understand that the European winters are generally tempered by winds passing over Atlantic waters that have been warmed by the Gulf Stream currents, while much of the winter weather in Virginia originates over the colder landmass of North America. The reverse holds true in summer: summer winds in England are kept relatively cool by the Atlantic waters while summers in the Middle Atlantic are often accompanied by heat ushered in from air that has moved over a sun-baked land mass.

As the colonies became established and expanded throughout Virginia and Maryland, colonists documented the weather in diaries, letters and personal journals. Most weather data was recorded in this fashion for over 250 years, roughly until the middle of the 19th century, when the advent of the telegraph enabled government organizations to start tracking and forecasting the weather. The U.S. Army began documenting the weather as early as 1819.

Examination of the colonial weather records reveal that the weather of the Middle Atlantic region has not changed significantly over the past few centuries. Overall, temperatures appear to have warmed slightly, but snowfall and rainfall patterns have remained the same. Also, the Middle Atlantic region has experienced wide variations of temperature and weather within a season, and from year-to-year, even in colonial times.

The following provides a general weather comparison of the Middle Atlantic region over the past three centuries:

Snow: The snowfall average for Washington has remained relatively constant. For example, the average annual snowfall for the city during the early 1800's was about 15 inches, slightly less than the 17 to 18-inch average of today. There were some years when very little snow fell, and other years when snowfall was very heavy. During the winter of 1805-1806, only 4.5 inches of snow fell – similar to the winter of 2000-2001.

However, during the winter of 1771-1772, over 50 inches of snow fell – much like the snowy winter of 1995-1996 when 40 to 60 inches of snow fell on the Washington area. The colonists would describe very heavy snowfalls as quite unusual and very disruptive.

Cold: Frigid weather, which coats the Chesapeake Bay and Potomac River with ice for weeks at a time, has been fairly uncommon. In colonial records, an extended freeze would be described as if the Indians had rarely seen such an event. Also, early records show that sub-zero temperatures remained somewhat rare. However, a few of the colonial winters were exceptionally cold. The Freeze of 1779-1780 may have been the most extreme freeze in recorded history. Most of the Chesapeake Bay and all of the area's rivers were frozen, and huge ice mounds accumulated on nearby Atlantic beaches. Recently, there have been severe cold spells that have frozen the Chesapeake Bay and Potomac River, most notably during the winter of 1976-1977.

Heat: Colonial summers in the Middle Atlantic proved very hot and humid. The colonists complained frequently about the stifling heat.

They would sometimes cool off in their icehouses (rooms sunk into the ground that were filled with ice and snow from the previous winter). Temperatures of 100 degrees were recorded in colonial times, but did not occur often. Today, summer weather remains much like it was 300 years ago.

Severe Weather: Severe thunderstorms and tornadoes have been observed since colonial times. Lightning was documented to have killed some of the first colonists in Jamestown, and early writings indicate that the colonists issued warnings against standing near tall trees and stone fireplaces during thunderstorms. Tornadoes were documented to have occurred in the Middle Atlantic as early as the 17th century.

Tropical Storms and Hurricanes: Tropical storms and hurricanes have impacted life and travel in the Middle Atlantic since the beginning of colonial records. Perhaps the worst hurricane to hit Virginia occurred in the year of 1667, when the region was devastated by high winds and tidal flooding. More recently, the Washington area has been hit by many strong tropical systems, including the storms of 1896, 1933, 1954 and 1972.

John Smith's Accounts of Virginia's Weather, 1607

Captain John Smith wrote one of the first accounts of the weather in the Middle Atlantic region. He described the climate of Virginia soon after he had settled in Jamestown in 1607:

"The sommer is hot as in Spaine; the winter colde as in Fraunce or England. The heat of sommer is in June, July and August, but commonly the coole Breezes asswage the vehemencie of the heat. The chiefe of winter is halfe December, January, February and halfe March. The colde is extreme sharpe, but here the Proverbe is true, that no extreme long continueth."

Captain John Smith documented the weather of Virginia while he established the colony of Jamestown in 1607-1608. The English colonists were very curious about the climate of the New World and wrote numerous descriptions about Virginia's weather.

"The winds here are variable, but the like thunder and lightning to purifie the aire, I have seldome either seen or heard in Europe. From the southwest come greatest gusts with thunder and heat. The northwest winde is commonly coole and brings faire weather with it. From the North is the greatest cold, and from the East and Southeast as from the Burmudas, fogs and raines. Sometimes there are great droughts, other times much raine, yet great necessitie of neither..."

The Native American's Five Seasons, 1608

The colonists of Jamestown discovered that the natives of Virginia had divided the year into seasons that were similar to their own. However, they had divided fall into two seasons. Below were the five seasons of the native Virginians:

- *Popanow* was the winter season;
- *Cattapeuk* was the spring season;
- *Cohattayough* was the summer season;
- *Nepinnough* was the early fall season when corn was eared; and
- *Taquitock* was the late fall season when the leaves would fall.

The fall seasons were also the "Seasons of Feasts" for the native Americans. The feasts, which started in September and lasted until the middle of November, were a time when fruits and vegetables were ripe and abundant, and the "fish, fowl and wild beasts exceedingly fat."

John Clayton describes Virginia's Weather, 1685

John Clayton, Rector of Crofton in Yorkshire, left England for the purpose of making scientific observations in the New World. In 1685, he landed in Jamestown and began documenting the weather of eastern Virginia. Below are excerpts from his writings:

Winter - *Their Winter is a fine clear Air, and dry, which renders it very pleasant: Their frosts are short, but sometimes very sharp, that it will freeze the Rivers over Three miles broad... Snow falls sometimes in Pretty Quantity, but rarely continues there above a Day or two.*

Summer - *The Air becomes stagnant that the Heat is violent and troublesome. In September the Weather usually breaks suddenly, and there falls generally very considerable Rains.*

Lightning - *'Tis incredible to tell how it will strike large Oaks, shatter and shiver them, sometimes twisting round a Tree, sometimes as if it struck the Tree backwards and forewards.*

Tornadoes or Dust Devils - *There be frequent little sorts of whirl-winds, sometimes two or three yards, sometimes forty, which whisking round in a Circle, pass along the Earth, according to the Motion of the Cloud, from whence they issue: as they pass along with their gyrous or circular motion, they carry aloft the dry leaves into the Air, which fall again often in places far remote.*

Benjamin Franklin Documents Coastal Storm Tracks, 1747

Benjamin Franklin is credited as the first to understand and document the movement of coastal storms along the East Coast of North America. These storms, now often referred to as *Nor'easters*, move up the East Coast from southwest to northeast and are usually accompanied by strong northeast winds. Early thinking was that the storms would move in the same direction as their surface winds – thus the storms were thought to move down the coast, from the northeast to the southwest. Franklin realized this theory was erroneous and wrote a letter on July 16, 1747 which described his thoughts:

We have frequently, along this North American Coast storms from the Northeast which flow violently sometimes three or four days. Of these I have had a very singular opinion some years that though the Course of the Wind is from the northeast to southwest yet the course of the storm is from the southwest to northeast; that is the air is in violent motion in Virginia before it moves in Connecticut and in Connecticut before it moves at Cape Sable, etc.

Franklin used a severe coastal storm which occurred during the fall of 1743 as an example to prove that storms move up the coast. On the evening of October 21, 1743, Franklin had hoped to observe a lunar eclipse in Philadelphia. However, the lunar eclipse was hidden by cloud cover associated with a fierce coastal storm which caused damage along the Virginia coastline. Franklin later learned from sources in Boston that the same lunar eclipse was in full view in the Northeast. The day after the eclipse, a violent storm moved into New England causing significant damage along the Northeast coastline. The lunar eclipse set a reference point in time which proved to Franklin the same storm had hit both the Middle Atlantic region and New England. The storm had moved northeast up the coast, taking about a day to move from Virginia to New England.

Thomas Jefferson on Virginia's Climate, 1781

Thomas Jefferson had an active, scientific interest in the weather. He recorded the weather from both Williamsburg and Monticello, wrote papers and letters describing the weather and climate, and planned the first network of weather observers. He carefully documented temperatures, winds, and precipitation, and also described weather trends and patterns. In 1781, Jefferson wrote *Notes on the State of Virginia* in which he discussed Virginia's climate along with many other topics.

Jefferson noted that temperatures were generally colder in the mountains than along the Atlantic coast. He supported his findings by noting that the plant and animal life at high elevations more closely resembled the life in re-

gions far to the north with colder climates. He also noted that winds in the mountains prevail from the northwest, while winds along the coast tend to prevail from the northeast.

Many documents from Jefferson include references to the weather. One particularly interesting letter from Jefferson was written after he had experienced the harsh winter weather of Vermont. The letter was addressed to his daughter and reads as follows:

"On the whole, I find nothing anywhere else, in point of climate, which Virginia need envy to any part of the world. When we consider how much climate contributes to the happiness of our conditions, by the fine sensation it excites, and

Thomas Jefferson took a scientific interest in the weather. Jefferson documented the weather and climate of Virginia in journals, letters, and publications. *Library of Congress*

EARLY THERMOMETERS

In 1778, Thomas Jefferson wrote to a friend in France, "Fahrenheit's thermometer is the only one in use with us. I make my daily observations as early as possible in the morning and again about 4 o'clock in the afternoon, these generally showing the maxima of cold and heat in the course of 24 hours."

Jefferson's mention of *Fahrenheit's thermometer* was a reference to the mercury thermometer which was invented by Gabriel Fahrenheit in 1714, the predecessor of the modern thermometer. Fahrenheit also introduced the temperature scale that still bears his name. Fahrenheit assigned 0 to be the lowest at which he could get saturated salt water to freeze. He then assigned 100 to be the human body temperature. Due to inaccuracies, body temperature was later found to be 98.6. With the Fahrenheit scale, pure water freezes at 32 and boils at 212 (at sea level).

In 1742, Anders Celsius introduced the *Centigrade scale*, based on 0 for freezing and 100 for boiling of pure water at sea level. Europeans began standardizing on the Centigrade scale by the early 1800's. In 1948, the term *Celsius* was officially adopted in place of *Centigrade*.

Before Fahrenheit's mercury thermometer, gas, water, and alcohol thermometers were in use, employing various temperature scales. Galileo Galilei invented a gas thermometer in 1593, and in the 1630's, liquid-in-glass thermometers were introduced.

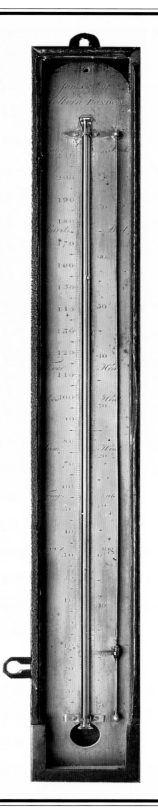

Thomas Jefferson's thermometer. Jefferson recorded the temperature daily and kept a detailed weather journal for many decades. This thermometer hangs on a wall at Monticello.
Thomas Jefferson Foundation and Monticello

the productions it is parent of, we have reason to value highly the accident of birth in such a one as that of Virginia."

"Indian Summer" Weather of 1783

The first documented use of the term "Indian Summer" occurred in an East Coast weather diary during one of the coldest winters of the 18th century. The winter of 1783-84 began early, with snow falling in mid-November and continuing into early December. On December 8, mild weather set in and by December 11 all of the snow on the ground had melted. The diary entry on December 8 described the mild weather as "Indian Summer." Today, "Indian Summer" refers to mild fall weather after the season's first frost has occurred.

The Little Ice Age

While the colonies were getting established in the New World, much of the Northern Hemisphere was experiencing the Little Ice Age. This was a colder-than-average period that started prior to 1500 and continued until the early 1800's.

The Little Ice Age is not what we consider to be an actual ice age of widespread glacial advances and dramatic climate cooling, but rather a relatively short period of time that had numerous cold winters intermingled with a few mild winters, and many cooler than average summers. Based on European weather records, the two coldest periods were from 1570 to 1600 and from 1690 to 1740. During those years, glaciers in the Alps grew larger. The Little Ice Age lasted 300-500 years, which is considered a very short period of time by climatological standards.

An illustration of a Frost Fair held on the ice of London's Thames River during the 1600's. Cold weather occurred from the late 1500's through the early 1800's, and this period is called the "Little Ice Age." Both England and the Colonies had unusually cold winters during the 1600's and 1700's.

In Europe, life changed drastically during the Little Ice Age. Farms were abandoned as crops failed; hunger and suffering became widespread, especially among the poor; and Eskimos were seen as far south as Scotland, due to a southward push of sea ice.

As the colonists were struggling through their first winter in 1607, the Little Ice Age was in full force. England experienced the same severe winter cold that was felt in the colonies. The Thames River froze solid during the winter of 1607 and "frost fairs" began to take place on the river's ice. The frost fairs continued into the 1800's, whenever the Thames River would freeze significantly. Today, the Thames River rarely freezes.

Although the temperatures during this period were below what is considered average today, there were still winters which were quite mild and free from heavy snow and ice. The best example occurred during the late 1700's, when there was a stretch of mild winters which sparked discussions of warming climates.

Global warming in the late 1700's?

During the late 1700's, a series of mild winters caused many to speculate that the climate was warming. The first mention of a warming climate was made in 1770 by Dr. Hugh Williamson, who said that winters were getting progressively warmer. He presented a paper to the American Philosophical Society discussing the change of climate in the Middle Colonies. Later in the century, Professor Samuel Williams, an early contributor to American meteorology, stated that "the winter is less severe, cold weather does not come on so soon." In 1781, Thomas Jefferson addressed the subject of climate change in his *Notes on the State of Virginia.* Jefferson wrote:

Both heats and colds are becoming much more moderate within the memory even of the middle-aged. Snows are less frequent and less deep. They do not often lay, below the mountains, more than one, two, or three days, and very rarely a week. They are remembered to have been formerly frequent, deep, and of long continuance. The elderly inform me the earth used to be covered with snow about three months every year. The rivers, which then seldom failed to freeze over in the course of the winter, scarcely ever do so now.

Noah Webster theorized about the warming weather of the late 1700's, suggesting that it was based on agriculture. He explained in a paper to the Connecticut Academy of Arts and Sciences that the clearing of forests and cultivation of land exposes the ground to direct sunlight, causing greater heating of the land and air. He went on to further explain that the weather was becoming more variable, as well as generally warmer, as forests disappeared.

During the 1920's, speculation of a warming climate became a topic of conversation and debate once again. Robert Ward, Professor of Climatology at Harvard University, researched colonial winters in New England to look for weather trends. In the colonial records he found ample evidence of both severe, stormy winters and mild, pleasant winters. In 1925, Professor Ward wrote: "A mild winter with light snowfall is just as old-fashioned as one with severe cold and heavy snowfall. There were plenty of both kinds of winters in the past, and there will be plenty of both kinds in the future."

EARLY WEATHER EVENTS

The Storm of 1586 that Ended Raleigh's First Colony

In the spring of 1584, the well-known adventurer and English political figure named Walter Raleigh (also spelled Ralegh) sent a surveying party to find a site for a colony along the coast of the New World. They found a good location at Roanoke Island, located on the outer banks of what is now North Carolina. The climate seemed temperate, and the land was abundant with resources. Several months later, the exploration party returned to England with favorable reports. Raleigh, excited about the prospects of a colony, named the land "Virginia" in honor of Elizabeth, the Virgin Queen of England. The Queen, in return for the honor, knighted him Sir Walter, Lord and Governor of Virginia. However, she refused his request to fund a national enterprise to colonize Roanoke Island.

Undaunted, Raleigh managed to fund a small flotilla and a contingent of 600 men to settle Roanoke Island. They set sail from England on April 9, 1585, and arrived on Roanoke Island late in the summer of 1585. In the fall, the colony sent glowing

reports of the New World back to England, by way of a returning ship:

...It is the goodliest and most pleasing territory of the world, and the climate so wholesome that we had not one sick since we touched land here...

As winter set in, however, harsh weather and trouble with the Indians began to plague the

Sir Walter Raleigh funded two colonies at Roanoke Island. The first colony failed after the colonists experienced difficult winter weather, trouble with the natives, and a strong spring storm that almost destroyed their supply ships. Raleigh's second colony on Roanoke Island became the fabled "lost colony" after all of the colonists mysteriously disappeared.

colony. The colonists survived the winter, but complained of many hardships. By spring, most colonists were ready to return to England. Sir Francis Drake arrived in the colony on June 10, 1586, hoping to re-supply and reinvigorate the colony. As Drake's ships lay at anchor off the coast, a ferocious storm hit the region. The storm was described as "extraordinary" and lasted three days. During the storm, hail the size of "hen eggs" pelted the colony. Waterspouts were also observed. Drake's ship, the *Primrose*, broke its 250-pound anchor and many supplies were lost. It was remarked that the storm caused more damage to Drake's fleet of 23 ships than did all of their previous sea battles with Spain.

The fierce storm further strengthened the colonists' resolve to return to England. They felt that the storm was a sign from God that they should return home. When the storm had passed, the colonists boarded Drake's ships and set sail for England. A year later, Raleigh would start another colony on Roanoke Island – this colony would become the fabled "lost colony" after all of the colonists mysteriously disappeared.

Jamestown's First Summer of 1607

On December 20, 1606, three merchant ships from the Virginia Company set sail from England to establish a colony near the Chesapeake Bay. The ships carried 104 men who were hoping to find gold and to locate a water route to the Pacific. On May 13, 1607, the colonists landed on Jamestown Island, located on the James River near the mouth of the Chesapeake Bay. Soon after arriving, the colonists began to encounter hardships – most notably from attacks by the Algonquian Indians, but also from the very hot, humid Virginia weather. John Smith wrote:

Our extreme toil in bearing and planting pallisades, so strained and bruised us, and our continual labor in the extremity of heat had so weakened us, as were cause sufficient to have made us as miserable in our native country as any other place in the world.

As the hot summer progressed, many colo-nists fell ill. Thirst drove the colonists to drink from the nearby marsh and briny river, which may have contributed to their sickness and distress. The colonists perished in epidemic proportions during the long, hot summer, with most succumbing to disease; but the Algonquian attacks also accounted for some of the casualties.

After surviving the first summer in Jamestown, Captain John Smith and others believed that the colonists required a period of "seasoning" to the climate of Virginia to thrive in the New World. In actuality, survival in Jamestown was more a matter of enduring the poor drinking water and dodging the Algonquian attacks than acclimating to the weather. A few years later, the concept of survival by "seasoning" was forgotten.

Jamestown's "Great Frost" and "Starving Time" Winters, 1607-1610

As the colonists of Jamestown were preparing for their first winter in 1607, fire destroyed much of their stores and supplies. In December of 1607, John Smith led a search party into Indian lands to search for food. At this time, extremely cold weather was settling into Virginia. John Smith wrote:

In the year 1607 was an extraordinary frost in most of Europe, and this frost was found as extreme in Virginia.

The European cold spell that Smith referenced represented one of the coldest winters ever experienced in England. During that winter, the Thames River in England froze, and London's first "Frost Fair" was held on the ice. The lone surviving diary from Jamestown's first year blamed the extreme cold for the deaths of half the colonists during the winter of 1607.

Two years later, the winter of 1609-1610 brought extreme cold back to Jamestown. The cold weather, combined with a lack of food, took a terrible toll on the colonists. Of the 500 colonists in Jamestown during the fall of 1609, only 60 survived the winter. The surviving colonists named the winter of 1609-1610 the "Starving Time."

No detailed weather records exist from the first years of Jamestown. However, in later manuscripts, colonists described succeeding winters as being more mild and pleasant in comparison to their first winters in Jamestown. Not long after the Starving Time Winter, the English Colonial Governor of Virginia, Sir Thomas Gates, wrote, "The winters are so mild that the cattle can get their food abroad, and swine can be fatted on wild fruits."

The "Dreadful Hurricane" of 1667

The Hurricane of 1667 is considered one of the most severe hurricanes to hit Virginia. The storm moved across the Lesser Antilles where it devastated St. Christopher and then tracked northwest and struck the Outer Banks of North Carolina. After making landfall, the storm turned northward and moved up the East Coast. The coastal regions of the Middle Atlantic were hit particularly hard. Several accounts attest to the fury of this great storm. The following account was published in London shortly after the storm:

Sir having this opportunity, I cannot but acquaint you with the relation of a very strange tempest which hath been in these parts which had began August 27th and continued with such violence, that it overturned many houses, burying in the ruins much goods and many people, beating to the ground such as were any waves employed in the fields, blowing many cattle that were near the sea or rivers, into them, whereby unknown numbers have perished, to the great afflication of all people, few having escaped who have not suffered in their persons or estates, much corn was blown away, and great quantities of tobacco have been lost, to the great damage of many, and utter undoing of others. Neither did it end here, but the trees were torn up by the roots, and in many places whole woods blown down so that they cannot go from plantation to plantation. The sea swelled twelve feet above its usual height drowning the whole country before it, with many of the inhabitants, their cattle and goods, the rest being forced to save themselves in the mountains nearest adjoining, while they were forced to remain many days together in great want. The tempest, for the time, was so furious, that it hath made a general desolation, overturning many plantations, so that there was nothing that could stand its fury.

The following is a letter from Secretary Thomas Ludwill to Lord Berkeley regarding the hurricane:

Jamestown Colony - this poor country is now reduced to a very miserable condition by a continental course of misfortune. On the 27th of August followed the most dreadful Hurry Cane that ever the Colony groaned under. It lasted 24 hours, began at North East and went around northerly till it came to west and so it came to Southeast where it ceased. It was accompanied with a most violent rain but no thunder. The night of it was the most dismal time I ever knew or heard of, for the wind and rain raised so confused a noise, mixed with the continued cracks of failing houses. The waves were impetuously beaten against the shores and by that violence forced and as it were crowded into all creeks, rivers and bays to that prodigious height that it hazarded the drowning of many people who lived not in sight of the rivers, yet were then forced to climb to the top of their houses to keep themselves above water. The waves carried all the foundations of the Fort at Point Comfort into the river and most of furnished and garrison with it...but

then morning came and the sun risen it would have comforted us after such a night, had it not lighted to us the ruins of our plantations, of which I think not one escaped. The nearest computation is at least 10,000 houses blown down, all the Indian grain laid flat on the ground, all the tobacco in the fields torn to pieces and most of that which was in the houses perished with them. The fences about the cornfields were either blown down or beaten to the ground by trees which fell upon them.

The storm passed inland, northeast of Jamestown, into northern Virginia. The huge tidal surge that flooded Virginia's coastal regions attests to this being one of the larger hurricanes to have hit the Middle Atlantic. Another hurricane or tropical disturbance may have passed very close to the Virginia coastline on September 10, since the "Dreadful Hurricane" of 1667 was followed by many days of rain.

Benjamin Franklin's Lightning Experiment, June 1752

Benjamin Franklin was fascinated by electricity and conducted many experiments with electrical current. His most famous experiment involved flying a kite during a thunderstorm with a key tied to the kite string. No one is certain that this experiment was actually conducted, and it could have easily produced fatal results to Franklin and nearby observers. However, an account of the experiment was written and published years later, presumably with input from Franklin. An excerpt reads as follows:

The kite being raised, a considerable time elapsed before there was any appearance of its being electrified. One very promising cloud had passed over it without any effect; when, at length, just as he was beginning to despair of his contrivance, he observed some lose threads of the hempen string to stand erect, and to avoid one another, just as if they had been suspended on a common conductor. Struck with this promising appearance, he immediately presented his knuckle to the key, and the discovery was complete. He perceived a very evident electric spark.

Where & how — my time is Spent

Jan: 26. At home all day alone
that is with the family. —

27. At home by ourselves the
day being dreadfully bad —

28. Just such a day as the former &
at home alone —

29. With much difficulty rid as
far as the mill the snow being
up to the breast of a tall horse
everywhere —

30. At home all day it being
almost impracticable to get
out. —

31. Still at home for the causes
above

Entries from George Washington's diary noting the "Washington and Jefferson" Snowstorm of January 26-31, 1772. The storm produced very heavy snow across the Middle Atlantic region, with both Thomas Jefferson and George Washington recording three feet of snow at Monticello and Mount Vernon. Washington's diary entries are as follows:

January 26. At home all day alone. That is with the family.
January 27. At home by ourselves the day being dreadfully bad.
January 28. Just such a day as the former and at home alone.
January 29. With much difficulty rid as far as the mill the snow being to the breast of a tall horse everywhere.
January 30. At home all day it being almost impracticable to get out.
January 31. Still at home for the causes above.

"The Great Fresh" of 1771

Ten days of rain in the Highlands of Virginia during May 1771 created a legendary flood of the James River in Richmond. The event was called "The Great Fresh" due to the tremendous surge of freshwater that flowed down the James River into the salty, tidal waters of eastern Virginia. Historical accounts state that the James River rose an incredible 40 to 45 feet above normal water levels. Houses, gristmills and tobacco warehouses were swept into the river and washed downstream. Cries of help could be heard coming from houses that had been washed into the river. Over 150 people lost their lives in the flood.

The rain began in the Blue Ridge Mountains in May 1771, while clear skies were observed in the tidewater regions of Virginia. As the rain continued for many days in the mountains, the James River began to rise in Richmond. For 60 hours, river levels rose – increasing as much as 16 inches in one hour. At a port just below Richmond, ships were tied to trees as the James River flooded. As the river continued to rise, the ropes holding the ships were tied higher and higher up in the trees. At the crest of the flood, the ships were tied to the tops of trees. Soundings made by the ships showed that they were floating 40 feet above the normal tide levels.

Extensive flooding was noted on many rivers, including the Potomac, Rappahannock and Rivanna Rivers. On the Rivanna River, Thomas Jefferson lost his gristmill to the floodwaters. On the Potomac River, at the port of Dumfries, a large portion of the 1770 tobacco crop was lost when storehouses were flooded. In the fields throughout eastern Virginia, over half of the tobacco seedlings were lost. The total loss of tobacco was estimated at over 3 million pounds.

Since tobacco exports were the main source of income for the colony, the flood was financially devastating. In July, the Virginia General Assembly convened three months early to discuss the "Melancholy Catastrophe" and plan relief for the tobacco farmers. They decided to raise relief money by placing a new tax on taverns, tobacco and wagon vehicles, excluding only farm wagons. The lawmakers described the tax as "easy to the people."

The Washington and Jefferson Snowstorm of 1772

The diaries of both George Washington and Thomas Jefferson reference a great snowstorm in January of 1772 that is one of the largest documented snowfall events for the piedmont regions of Virginia and Maryland. Snow piled up three feet deep from the Blue Ridge Mountains of Virginia east to the Chesapeake Bay. Even west of the Blue Ridge in Winchester, Virginia, the snow was measured at 2 feet, 9 inches. However, farther to the north, towards Philadelphia, the snowfall diminished rapidly.

The storm took shape on January 26, 1772, in the southeastern states. As the storm developed, a raw northerly wind began to blow and the sky filled with clouds. Snow began to fall during the evening of January 26, and by the morning of January 27 six inches of snow had accumulated at Mount Vernon. The snow continued to fall on January 27 with a stiff, northerly wind. Snow fell without abating until January 29 when there was a break, but the snow began again that evening.

When the storm ended on January 30, the snow measured three feet at both Monticello and Mount Vernon. The Gazette of Annapolis commented: "it is with utmost difficulty people pass from one house to another." George Washington wrote in his personal diary "the snow being to the breast of a tall horse everywhere," and Thomas Jefferson wrote, "The deepest snow we have ever seen." The snow caused the postponement of the meeting of the General Assembly in the colonial capitol of Williamsburg and caused mail messengers in Virginia to be delayed for many days.

When the storm began, Thomas Jefferson was returning home from his honeymoon with his new bride. The newlyweds made it to within eight miles of Monticello before having to abandon their carriage in the deep snow. Both finished the ride on horseback, traversing the difficult mountain roads in the blinding snow. The newlyweds arrived home late on the night of January 26, exhausted from the trip. Both would recount years later about their tiresome journey in the snow and most notably about the

Thomas Jefferson's weather records for December 25, 1776. Jefferson's records show that 21 inches of snow fell at Monticello on Christmas night, 1776. The high temperature was 30°F and the low temperature was 23.5°F. Farther to the north, up to 30 inches of snow fell in northern Virginia and central Maryland. *Library of Congress*

"dreariness" of arriving to a cold, dark home during a snowstorm.

The cold and stormy pattern returned to the Middle Atlantic region in March of 1772. Three storms – on March 11, March 17 and March 20 – laid down a blanket of snow that totaled 20 inches in Central Maryland. All told, the snowfall totals for the winter of 1771-1772 would have been over 50 inches – comparable to the recent winter of 1995-1996.

The Christmas Snowstorm of 1776

On Christmas day in 1776, a heavy snowstorm began to lash the East Coast from North Carolina to New York. This storm is well known because it impacted General Washington's Continental Army as it crossed the Delaware River on Christmas night, en route to Trenton, New Jersey, to attack the Hessian forces that occupied the city. The storm dumped about two feet of snow from Central Virginia to Central Maryland in a twenty-four hour period. However, farther to the northeast in New Jersey, much less snow fell due to a changeover to sleet and freezing rain.

The storm was in full fury on Christmas night when Washington's army crossed the Delaware River. At the time of the crossing, the snow was in the process of changing to sleet. For the rest of the night, the precipitation alternated between snow, sleet and freezing rain. A diary of a member of Washington's staff briefly describes General Washington and the Delaware crossing:

"He stands on the bank of the stream, wrapped in his cloak, superintending the landing of his troops. He is calm and collected, but very determined. The storm is changing to sleet and cuts like a knife."

George Washington and his Continental Army crossed the ice-clogged Delaware River on Christmas night of 1776 in a snowstorm. Washington and his army were en route to Trenton, New Jersey to attack Hessian forces. Leutze's painting of the crossing does not accurately portray the poor visibility caused by the blowing snow and sleet. One foot of snow and sleet fell in Trenton, New Jersey, and two feet of snow fell from Northern Virginia to Central Maryland.

A LOOK BACK AT 1797

Colonial and pre-Civil War weather records for the Middle Atlantic region are typically found in period journals, letters and articles. Most weather data was documented in this fashion until the middle of the 19th century, when the advent of the telegraph enabled government organizations to start tracking and forecasting the weather.

Researching and understanding early weather events often involve deciphering crude, hand-written weather descriptions on the yellowing pages of personal journals. **"Journal of Weather kept at the City of Washington"** offers a good example of such documentation. The manuscript chronicles the daily weather conditions in Washington during a one-year period, starting June 1797. The journal describes the weather in general terms of everyday life, such as describing the severity of the cold by the number of blankets needed at night.

In addition to the weather descriptions, the **Journal** details where and when ice formed on the Potomac River. Ice on the river was watched closely because it impacted shipping to Washington. In addition, the river's ice was harvested every year. Ice would be cut into blocks and stored in icehouses for use in the spring and summer months, when it was used primarily to cool drinks. A good ice harvest would mean ice for much of the year. The journal noted that ice formed briefly as far south as Alexandria in early December. However, the ice harvest for the winter was poor — the winter's warm, rainy weather kept the Potomac River relatively ice-free.

For the winter of 1797-98, there were only three small snowfalls, with the largest snow occurring on January 23, which accumulated three to four inches. The first frost occurred on October 12. Spring arrived early, with very mild temperatures and thunderstorms occurring in early March.

The journal also includes frequent descriptions about the cool breezes that followed summertime thunderstorms. For Washington residents of the 18th century, who lived without the luxury of air conditioning, the cool breezes of a thunderstorm were a very welcome relief from the hot, humid weather. The summer of 1797 featured many thunderstorms, which helped break a fairly serious springtime drought that had jeopardized the year's corn crop. The maximum temperature for the summer was 95°F, which occurred on June 17.

Thomas Jefferson also documented the weather of 1797 from his home in Monticello. Jefferson noted that the first frost at Monticello occurred on November 2 and that there were only two small snowfalls, adding up to less than two inches. In a letter to James Madison, Jefferson wrote of flooding rains that occurred in January 1797, which prevented mail from arriving in Charlottesville.

Overall, the weather of 1797 was quite uneventful. The year was fairly mild and free from severe storms and heavy snowfall. The most interesting feature of the journal data is that it paints a picture of the Washington-area weather 200 years ago, which closely resembles its present-day weather. The biggest difference between 1797 and today is that we now have technology like air conditioning and ice-makers that help make living in Washington's changeable weather much more comfortable.

*The author of **"Journal of Weather kept at the City of Washington"** is unknown and no other volumes are known to exist. The journal was purchased by the United States Weather Bureau in 1909 and is currently kept at the National Oceanic and Atmospheric Administration (NOAA) Library in Silver Spring, Maryland.*

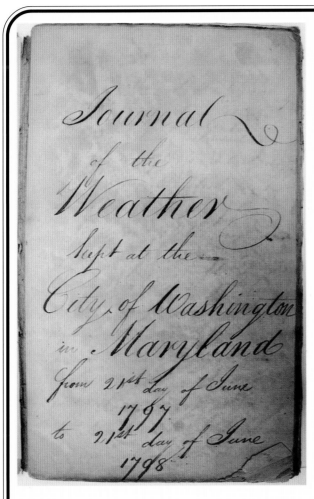

The cover page for the *Journal of Weather kept at the City of Washington in Maryland from 21st day of June 1797 to 21st day of June 1798.* The manuscript chronicles the daily weather conditions in Washington during a one-year period, starting June, 1797. The journal describes the weather in terms of everyday life, such as describing the severity of the cold by the number of blankets needed to keep warm at night.
NOAA Library

The cover of the first edition of the Hagers-Town Almanack, 1797. The Hagers-Town Almanack was established in 1797 by John Gruber and is the second oldest almanac being published in the U.S. (The Old Farmer's Almanac, established in 1792, is the oldest almanac.) The first editions of the Hagers-Town Almanack were printed in German, with English editions added in 1822. The 1797 Almanack included a table to foretell the weather based upon the position and stage of the moon, the season, and wind direction.   *NOAA Library*

After crossing the Delaware, the American soldiers marched nine miles to Trenton. This march is often described in history books because some of the soldiers did not have shoes and had wrapped rags around their feet. As the soldiers marched, there were trails of bloody footprints in the snow. A strong northeast wind made the march even more difficult as it whipped the freezing precipitation into the marching soldiers.

Upon reaching Trenton, Washington's army circled the town and engaged the Hessian soldiers. The heavy precipitation quickly rendered all flintlock rifles useless. The battle proceeded with bayonet, sword and artillery. Washington's artillery was able to maintain dry powder during the storm. The Americans prevailed during the battle and captured 900 soldiers, hundreds of muskets, and six pieces of artillery. Although the number of soldiers and weapons captured at Trenton was not significant, the impact on the morale of both the British and American armies was huge. The effort for American independence had been revitalized.

The precipitation in Northern Virginia and Central Maryland was mainly snow, with a total snowfall accumulation of 24 to 30 inches. In Central Virginia, Thomas Jefferson measured 21 inches of snow at Monticello. Closer to the coast, the precipitation was mixed with sleet and freezing rain and accumulations were reduced. Almost 12 inches of snow and sleet fell near Trenton, New Jersey. To the south, 4 inches of snow fell in central North Carolina.

The Cold Wave of 1780 that froze the Chesapeake Bay

An unprecedented cold wave occurred during the winter of 1779-1780 that froze all the waterways of the Middle Atlantic region, including the Potomac River and most of the Chesapeake Bay. The cold weather set in during December of 1779 and area rivers were frozen before Christmas. Continuous cold weather occurred during the month of January 1780 and lasted through the first week of February. Temperatures did not moderate until the second week of February, when the first thaw occurred.

The Virginia Gazette reported that most of the lower bay was iced over, almost to the Virginia Capes. On the northern part of the Bay, sleighs crossed from Baltimore to Annapolis and from Annapolis to the Eastern Shore. To the south, Norfolk, Hampton, Newport News and Portsmouth were connected by thick ice that supported foot traffic between the ports. The York River was crossed on foot near Gloucester and the James River could be crossed on the ice near Williamsburg. Even farther south, the Albemarle Sound was frozen thick enough to allow American soldiers to cross on foot. Thomas Jefferson noted that never before had the tidal waters frozen to such a large extent. Along the Atlantic beaches of the Delmarva, ice mounds piled up to a height of twenty feet – a result of the breaking waves and spray freezing on the beach. The ice mounds did not melt until spring.

Prolonged and repetitive cold air masses were responsible for the great freeze. The coldest weather occurred between January 6-8, January 13-16, and January 19-29. Zero-degree temperatures reached as far south as Williamsburg during the coldest spell. Although the first thaw set in during mid-February, Baltimore's harbor remained closed due to ice until March 9. The port at Philadelphia was locked in ice from December 21 to March 4 – about two and a half months! When the ice began to thaw in the spring of 1780, many wooden ships were destroyed by the thick, moving ice.

Snowbound in 1784

One of the coldest and snowiest winters in history can be linked to a volcanic eruption in Iceland. During the summer of 1783, an unusual blue haze filled the sky over North America. The blue haze was also seen over Europe, western Asia and North Africa. The haze was a result of a catastrophic volcanic eruption in Iceland that emitted enormous quantities of gases and volcanic ash into the atmosphere. The haze partially blocked the sun and led to colder than average temperatures. Benjamin Franklin correctly speculated that the cold weather was a result of the haze and the volcano.

The winter of 1784 was also beset by many heavy snowstorms that moved up the East Coast. The snowstorms began in November and continued through March. Deep snow covered the ground during much of the winter. George Washington wrote on March 5, 1784, "arrived at this Cottage on Christmas Eve, where I have been locked up ever since in frost and snow." James Madison wrote to Thomas Jefferson and noted that he had never seen a season so severe in any preceding winters. The Richmond Gazette reported that the deep snow prevented travel and trade in Virginia.

The ice on the Potomac River at Alexandria, Virginia, was solid from December until the middle of March. Ice caused the harbor at Baltimore, Maryland, to be closed from January 2 to March 25 – a total of twelve weeks. This was a longer extent than the winter of 1780 (by a couple of weeks). Snow and cold temperatures lasted through most of March.

The British army invaded Washington and set fire to the city on August 24, 1814. A day later, a line of severe thunderstorms spawned a tornado in Washington that killed several British soldiers and caused significant damage to the city. Heavy rain associated with the storm helped extinguish the fires that burned throughout Washington. *Washingtoniana Division, D.C. Public Library*

The Tornado and the Burning of Washington, August 25, 1814

During the summer of 1814, British warships sailed into the Chesapeake Bay and headed towards Washington. The warships sailed up the Patuxent River and anchored at Benedict, Maryland on August 19, 1814. Over 4,500 British soldiers landed and marched towards Washington. The British mission was to capture Washington and seek revenge for the burning of their British Capitol in Canada, for which they held the United States responsible.

A force of 7,000 Americans was hastily assembled near the Potomac River to defend Washington. During the afternoon of August 24, in 100°F heat, the two armies clashed. The British Army quickly routed the less disciplined American volunteers, mostly due to a series of American blunders and a new British rocket that did little damage, but unnerved the raw American troops with a very loud, shrill noise. President Madison and Secretary of State Monroe, who had led a group of officials to watch the battle, were almost captured in the confusion. It was noted that the 100°F temperatures added to everyone's discomfort.

After the battle, the British Army marched quickly into Washington while American soldiers, United States government officials, and residents fled the city. There were no officials left in Washington from whom the British could seek terms of surrender. The British admiral ate dinner in the White House, then gave the order to set fire to Washington. Within hours, the White House, the Capitol, and many other public buildings and residences were burning.

On the morning of August 25, Washington was still burning. Throughout the morning and early afternoon, the British soldiers continued to set fires and destroy ammunition supplies and defenses around the city. As the soldiers spread fire and destruction throughout the city, the early afternoon sky began to darken and lightning and thunder signaled the approach of a thunderstorm. As the storm neared the city, the winds began to increase dramatically and then built into a "frightening roar." A severe thunderstorm was bearing down on Washington, and with it was a tornado.

The tornado tore through the center of Washington and directly into the British occupation. Buildings were lifted off of their foundations and dashed to bits. Other buildings were blown down or lost their roofs. Feather beds were sucked out of homes and scattered about. Trees were uprooted, fences were blown down, and the heavy chain bridge across the Potomac River was buckled and rendered useless. A few British cannons were picked up by the winds and thrown through the air. The collapsing buildings and flying debris killed several British soldiers. Many of the soldiers did not have time to take cover from the winds and they laid face down in the streets. One account describes how a British officer on horseback did not dismount and the winds slammed both horse and rider violently to the ground.

The winds subsided quickly, but the rain fell in torrents for two hours. (There may have been a second thunderstorm that followed quickly after the first thunderstorm.) Fortunately, the heavy rain quenched most of the flames and prevented Washington from continuing to burn. After the storm, the British Army regrouped on Capitol Hill, still a bit shaken by the harsh weather. They decided to leave the city that evening. As the British troops were preparing to leave, a conversation was noted between the British Admiral and a Washington lady regarding the storm: The admiral exclaimed, "Great God, Madam! Is this the kind of storm to which you are accustomed in this infernal country?" The lady answered, "No, Sir, this is a special interposition of Providence to drive our enemies from our city." The admiral replied, "Not so Madam. It is rather to aid your enemies in the destruction of your city."

Hours later, the British forces left Washington and returned to their ships on the Patuxent River. The journey back was made difficult by the numerous downed trees that lay across the roads. The war ships that lay waiting for the British force had also encountered the fierce storm. Wind and waves had lashed at the ships and many had damaged riggings. Two vessels had broken free from their moorings and were blown ashore.

President Madison and other government officials returned to Washington and began the difficult process of setting up government in a city devastated by fire and wind. Never again would the British Army return to the city, and only rarely would Washington suffer damaging tornadoes.

"The Year without a Summer" in 1816

A series of volcanic eruptions in the early 1800's, culminating with the 1815 eruption of the volcano Tambora, in Indonesia, is thought to have caused a brief global cooling associated with "The Year without a Summer" in 1816. During the summer months of 1816, cold weather and frosts occurred in the Northeast and Middle Atlantic regions during multiple cold snaps. Crops were ruined and trees were damaged from North Carolina to New England.

The weather of 1816 began very cold. On January 9, 1816, Thomas Jefferson wrote from Monticello, "shivering and shrinking in body from the cold we now experience, my thermometer having been as low as 12°F this morning." Later in January, temperatures moderated and the rest of the winter averaged slightly above normal. Cold weather set in during March and lasted through May. April was particularly cold, with temperatures averaging almost 9°F below normal at Monticello. In New England, some trees did not bloom until the end of May. The weather briefly warmed during the first few days of June, then drastically cooled down on June 6. Temperatures plummeted from near 90°F in many eastern cities to the 30's and 40's. A day later, on June 7, a heavy snowfall occurred in the Northeast, with up to 10 inches of snow and 20-inch snowdrifts measured in Danville, Vermont. Upstate New York received 3 inches of snow and even Boston, Massachusetts observed a trace of snow. On June 10, a severe frost blackened fields of beans and cucumbers from Virginia to New England. In some areas of the Northeast, trees remained leafless well into June due to the long, cold spring.

Cold weather returned to the eastern U.S. during the first week of July, when another killing freeze occurred in the Northeast on July 5. In parts of New England and the Middle Atlantic, crop damage was severe and fruit trees were killed. In Savannah, Georgia, the temperature dropped into the 40's on July 4. In Pennsylvania, ice the thickness of window glass formed on the morning of July 5.

More cold descended upon the East on August 22, 1816, when a freeze was once again observed in New England. This time, frost extended as far south as western North Carolina. Crop damage was widespread and the growing season ended very early. Corn even froze on the stalk from the Middle Atlantic to New England. Also, significant tree damage was noted throughout the Northeast. The following year, in the spring of 1817, seed corn prices increased to

$4 a bushel (many times higher than the normal price), with much of the corn coming from the 1815 harvest.

An excellent description of "The Year without a Summer" is found in a weather journal that was written in Philadelphia, Pennsylvania by Charles Peirce. It is included below:

The Year, 1816. *The temperature of the whole year was only 49°F; it being the coldest year we have on record. Although there was no uncommonly cold weather during the three winter months, yet there was ice during every month in the year, not excepting June, July, and August. There was scarcely a vegetable came to perfection north and east of the Potomac. The cold weather during summer, not only extended through America, but throughout Europe. It was also the coldest summer ever known in the West Indies and in Africa.*

June, 1816. *The medium temperature of the month was only 64°F, and it was the coldest month of June we ever remember; there were not only severe frosts on several mornings, but on one morning there was said to be ice. Every green herb was killed, and vegetables of every description very much injured. All kinds of fruit had been previously destroyed, as not a month had passed without producing ice. From 6 to 10 inches of snow fell in various parts of Vermont; 3 inches in the interior of New York; and several inches in the interior of New Hampshire and Maine.*

July, 1816. *The medium or average temperature of this month was only 68°F, and it was a month of melancholy forebodings, as during every previous month since the year commenced, there were not only heavy frosts, but ice, so that very few vegetables came to perfection. It seemed as if the sun had lost his warm and cheering influences. One frosty night was succeeded by another, and thin ice formed in many exposed situations in the country. On the morning of the 5th there was ice as thick as window glass in Pennsylvania, New York, and through New England. Indian corn was chilled and withered, and the grass was so much killed by repeated frosts, that grazing cattle would scarcely eat it. Northerly winds prevailed a great part of the month; and when the wind changed to* *the west, and produced a pleasant day, it was a subject of congratulation by all. Very little rain fell during the month.*

August, 1816. *The medium temperature of this month was only 66°F, and such a cheerless, desponding, melancholy summer month, the oldest inhabitant never, perhaps, experienced. This poor month entered upon its duties so perfectly chilled, as to be unable to raise a warm, foggy morning, or cheerful sunny day. It commenced with a cold northeast rainstorm, and when it cleared the atmosphere was so chilled as to produce ice in many places half an inch thick. It froze the Indian corn, which was in the milk, so hard, that it rotted up on the stock, and farmers mowed it down and dried it for cattle fodder. Every green thing was destroyed, not only in this country, but in Europe. Newspapers received from England said: "It will be remembered by the present generation, that the year 1816 was a year in which there was no summer."*

Thomas Jefferson's weather records at Monticello showed that the summer of 1816 was very cool, over 4°F cooler than normal. However, there were no summertime frosts at Monticello. This could be partly due to its location on the top of a small mountain, which is not prone to early or late season frosts like valleys and low-lying areas. The coldest temperature re-

corded at Monticello during the summer of 1816 was 51°F, which occurred repetitively in June, July and August. Jefferson also noted that his icehouse held ice until October 11, 1816, which was about a month longer than normal, despite starting the spring with a relatively small supply of ice and snow.

Jefferson's records also showed that the summer of 1816 was very dry, with approximately 50% of the normal rainfall in June, July and August. The rainfall for June, July and August was 0.33 inches, 4.63 inches and 0.85 inches, respectively.

In 1906, the Maryland Weather Service studied temperature observations recorded near Baltimore during the summer of 1816. They concluded that the summer of 1816 was 8°F below normal in Baltimore. In their study, they wrote that the next coldest summer occurred in 1836, which was 4.3°F below normal.

The Great Snowstorm of January 1857

A very intense storm developed near the Gulf of Mexico on January 17, 1857, then tracked across Georgia and off the South Carolina Coast. The storm continued to intensify as it tracked northeast of the Virginia Capes. Heavy snow,

A common sleigh scene from the 1850's. On January 18, 1857, a severe snowstorm swept up the East Coast, dropping over 18 inches of snow from North Carolina to New England. Near-zero temperatures kept the precipitation mainly snow, even along the coast of southeast Virginia. Both Washington and Baltimore received 24 inches of snow.

Library of Congress

strong winds and cold temperatures were widespread along the entire Atlantic Seaboard. In the Washington area, snow began on the morning of January 18 and continued through January 19, accumulating 24 inches, with drifts up to 10 feet. In Norfolk, Virginia, snowdrifts were 20 feet high.

What made this storm unique was the extreme cold and very high winds. The entire Middle Atlantic region endured severe blizzard conditions with temperatures near 0°F. At Williamsburg, Virginia, the temperature was 3°F at the height of the snowstorm. The cold air behind the storm penetrated into Florida where the site of present day Miami had a temperature of 30°F.

The snowfall depths associated with the storm were very uniform. Heavy snow fell in a swath from interior South Carolina through Eastern Virginia and Maryland into New England. Snowfall totals were as follows:

Washington City	24 inches
Baltimore, Maryland	24 inches
Richmond, Virginia	24 inches
Gaston, North Carolina	22 inches
Providence, Rhode Island	18 inches
Portsmouth, Virginia	16 inches
Chapel Hill, North Carolina	15 inches
Stratford, Vermont	15 inches

Rarely does a snowstorm drop uniform amounts of snow from tidewater Virginia to the hills of Vermont. The frigid temperatures produced snow from the mountains to the coast. In addition, very strong winds were noted to be of hurricane force from eastern Virginia through eastern New England. Along the northeast coast, ships were blown ashore, church steeples and chimneys were blown down, and homes suffered wind damage.

Based on eyewitness accounts, it appeared that the storm had developed an eye, which passed over both Long Island and Boston. At Sag Harbor on Long Island, the winds and snow abated for 90 minutes and the sun broke through the clouds. After the lull, the snow and wind resumed. The same scenario happened at Boston for a period of almost two hours when the skies became partly cloudy during the middle of the storm. The lowest pressure readings at both sites occurred when the storm's eye passed overhead. At Sag Harbor, the lowest pressure was 28.91 inches, while at Boston, the lowest pressure was 29.24 inches. Washington was well to the west of the center of the storm and remained in the wind and snow for the duration of the storm.

The Snowstorm of January 1857 was a classic combination of unusually cold air, strong winds and heavy snow, which today would be termed a blizzard. However, the term "blizzard" was not yet in use for eastern snowstorms during 1857. Years later, the press would help bring the word "blizzard" into eastern weather vocabulary after the Blizzard of 1888 struck the Northeast.

The Thunderstorm and the Battle of Ox Hill (Chantilly), September 1, 1862

On a high hill just west of Fairfax, Virginia, a fierce Civil War battle was fought in a blinding thunderstorm on September 1, 1862. Two well-known Union generals were killed, and over two thousand casualties occurred in the dramatic fight that was greatly influenced by a line of thunderstorms associated with a strong cold front. The Battle of Ox Hill (or Chantilly) is also known as the only major Civil War battle to have been fought during a storm. To gain a full appreciation of the role that the weather played in the battle – as well as influencing the events that occurred later in the Civil War – the full context of the event needs to be explained:

In the closing days of August 1862, Union General John Pope's Army of Virginia suffered a defeat at the Battle of Second Manassas. The Union army, beaten but still intact, retreated to Centreville, Virginia and was heading back to the fortifications around Washington. General Lee, commander of the Confederate Army, hoped to stop the Union Army's retreat by circling around the Union forces and blocking their route. The location that General Lee had chosen to stop the Union Army was the Jermantown crossroad, located just west of Fairfax, Virginia.

Lee's strategy was to take Stonewall Jackson's Corps around the north side of the Union Army in a long, flanking move. To divert attention from Jackson's flanking move,

Longstreet's Confederate Corps stayed visible behind the Union Army. Meanwhile, Jeb Stuart's cavalry of 5000 rode ahead of Jackson's Corps to scout Pope's movements. Lee had planned for his armies to converge at the crossroad and block Pope's Union Army. As this maneuver began, there were cloudy skies with an increasing southerly wind. Clouds were noted to "race across the sky." Threatening weather was looming.

As Jackson's men were circling around the Union Army, Pope began to realize that a flanking move was underway by the Confederates, partly because Jeb Stuart had attacked a wagon train east of Chantilly. This alarmed Pope who sent Union General Issac Stevens and a small force of 4000 soldiers towards the Confederates. Stevens' forces surprised Jackson's men at Ox Hill, near the present day West Ox Road. Jackson abandoned his march towards the crossroads and went into a defensive position. At this time, the sky was darkening and the wind was blowing strong from the south, with flashes of lightning on the horizon

General Stevens, knowing he had surprised Jackson, made a courageous decision to immediately attack the Confederate brigades who were just starting to deploy into a defensive line. Stevens' first advance stalled in the face of a massive volley that caused numerous casualties among the Union ranks, including the general's son, Captain Hazard Stevens. The fighting began around 4:30 p.m., about the same time the thunderstorms were moving in.

General Stevens, distraught about having

In the rain, the Union Army retreats to Washington the day after the second Battle of Bull Run, August 31, 1862. On the next day, the Battle of Ox Hill occurred near Fairfax, Virginia, during a period of intense thunderstorms. The fighting raged for several hours in heavy rain and strong winds, ending in a stalemate. After the battle, the Union Army continued its retreat to Washington and the Confederate Army marched north, setting the stage for the Battle of Antietam.

seen his son fall in battle, decided to attack again while Jackson was still off balance. He grabbed a regimental flag from a wounded color bearer and personally led the charge. General Stevens was an easy target holding the battle flag and was quickly shot and killed. By now, the rain had started to fall heavily and began causing the soldier's black powder rifles to become unserviceable. With the rain in their favor, the outnumbered Union soldiers crashed through the center of the Confederate line. The Confederates, disorganized and confused, withdrew in considerable disorder. As the Confederates retreated, General Jubal Early's Brigade of Virginians arrived, filled the breach, and stopped the Union breakthrough from becoming a rout.

As the fighting continued, the sky became as dark as night and the thunderstorm lashed the soldiers with extremely heavy rain, strong winds and frequent lightning. The booms of cannons mixed with the loud explosions of thunder. The storm reduced visibility and the soldiers had a difficult time seeing their enemy. Sheets of rain blew horizontally across the battlefield, soaking the soldiers. Private Greely of the 19th Massachusetts recorded:

The roll of musketry and the roar of cannon left all of us unmoved, but the crash of thunder and the vividness of the lightning, whose blinding flashes seemed to be in our very midst, caused the uneasiness and disturbance among some of the bravest men.

The storm and rain continued as did the battle, and the intensity of the fighting did not diminish. Although the Confederates were caught off-guard by the Union attack, the hardened and tenacious Confederate veterans began to take control of the battle. The Union soldiers were slowly being driven back through the mud and rain. Fortunately, for the retreating Union troops, General Philip Kearney arrived with reinforcements. Hoping to reinvigorate the attack and urge the Union forces forward, Kearney pushed ahead of his men. With terrible visibility from the storm, General Kearny rode right up to the Confederate line. When he realized his mistake, he turned his horse and galloped through the thick mud back towards his division. The Confederates opened fire and Kearny was shot off of his galloping horse and was killed.

As the drenching rain continued, dry black powder cartridges became scarce and rifles misfired so often that the commanders told their troops to "give them the bayonet." A series of bayonet charges ensued. Finally, after over two hours of fierce hand-to-hand fighting, the heavy rain and cold temperatures began to dampen the will of the soldiers to fight. The Union army withdrew in the darkness and the Confederates held the field. Technically, it was a Confederate victory, but Lee had failed to accomplish his goal of stopping Pope's Army. The Union Army retreated to Washington, and the Confederates later turned north, setting the stage for the bloody Battle of Antietam.

The thunderstorms that occurred during the battle were associated with a strong, early-season cold front. The front was also accompanied by strong winds. Before the battle, the wind was strong from the south, recorded by the Naval Observatory in Washington to be at "Force 6."

(The Wind Force Scale ranged from 1 to 10 and was based on estimation.) The next day, on September 2, the Naval Observatory recorded winds from the northwest at "Force 4," and military records noted that northwest gales hampered shipping on the Potomac River. The Naval Observatory also recorded that 1.08 inches of rain fell during the storm of September 1, and they included the following remark: "Commenced an exceedingly heavy rain, with lightning and thunder, at 5:45 p.m."

The loss of Kearny and Stevens was a tremendous blow to the Union. Both men were popular and well-respected generals. In honor of Kearney, the U.S. Army began the tradition of awarding medals called the Kearny Cross for acts of courage. Soon afterwards, Congress authorized what is now known as the Congressional Medal of Honor. Hazard Stevens, General Stevens' son, recovered from his injuries suffered during the battle.

The Battle of Ox Hill was eventually forgotten, overshadowed by the Battle of Second Manassas and by Lee's invasion of the North. However, had the fierce thunderstorms not occurred, the Battle of Ox Hill may have turned out quite differently, potentially altering the course of the war. Without the storm, the Confederates had a good chance of stopping General Pope and making a move on Washington.

Currently, there are monuments to the Battle of Ox Hill at Fairfax Towne Center, which is located on the northeast side of the battlefield.

The "Mud March" Nor'easter, January 20-23, 1863

During late January of 1863, General Burnside's Union Army was in camp across the Rappahannock River from General Lee's Confederate troops, who were camped around Fredericksburg, Virginia. Burnside had been feeling increasing pressure to move against the Confederate Army after his defeat at the Battle of Fredericksburg, a month earlier, on December 13, 1862. Burnside's plan was to march his army several miles to the northwest of the Confederates and cross the Rappahannock, circling around the

The Army of the Potomac on their infamous "Mud March" during the Nor'easter of January 21, 1863. The Union Army became hopelessly bogged down in mud and aborted their march. Several days of heavy rain made the dirt roads impassible for the army's heavy supply wagons and artillery. Strong winds and temperatures in the 30's added to the miserable conditions for the soldiers.
Library of Congress.

left flank of Lee's army. He would then attack the Confederate Army near Fredericksburg. The weather had been fairly dry and mild for most of January and the prospects for a winter campaign seemed good.

On the morning of January 20, 1863, the Army of the Potomac formed columns and began the march up the Rappahannock River. Unknown to the soldiers, a massive storm was developing near the southeast coast and had started to move northward. Rain began falling during the evening

of January 20 and continued to fall heavily on January 21. Burnside's army quickly got bogged down in the mud. Temperatures hovered in the upper 30's, adding a chill to the drenched soldiers. Wagons sank to their wheel hubs in mud and artillery became hopelessly stuck. A team of 12 horses and 150 men could not pull one cannon out of the mud. Also, the soldiers slipped and fell repeatedly, while others lost their shoes in the thick mud.

General George Sykes of the Fifth Army Corps wrote a succinct summary of the storm:

On the night of the 20th, a violent storm of rain set in, making the roads impassable on account of the mud, rendering military movements impracticable. The entire command was turned out to repair and corduroy the roads.

By January 22, the rain had ended but the

entire Army of the Potomac was still mired in the mud. The weather remained cloudy and damp, with temperatures hovering in the upper 30's to near 40°F. The damp conditions and above-freezing temperatures kept the roads soft and muddy. Ammunition and supply wagons remained stuck fast, and horses and mules died of exhaustion in the mud. The challenge was no longer to cross the river and attack Lee, but instead to get unstuck from the mud and return to camp. Log roads were built over the mud with great effort, and the Union Army arrived back in camp on January 23. The campaign had been a dismal failure and General Burnside was even heckled by his own men as they marched back to camp in the mud.

A soldier's account gives vivid detail of the misery involved during the march:

For the clouds gathered all day thicker and darker, and night ushered in a storm of wind and pouring rain, harder for that moving army to encounter than a hundred thousand enemies; a driving rain that drenched and chilled the poor shelter less men and horses, and that poached the ground into mud deeper than the New England mind can conceive of, and stickier than – well, I am at a loss for a similitude. Pitch, for cohesion attraction, is but as sand compared with it.

Another account describes the muddy roads:

The rain lasted thirty hours without cessation. To understand the effect, one must have lived in Virginia through a winter. The roads are nothing but dirt roads. The mud is not simply on the surface, but penetrates the ground to a great depth. It appears as though the water, after passing through a first bed of clay, soaked into some kind of earth without any consistency. As soon as the hardened crust on the surface is softened, everything is buried in a sticky paste mixed with liquid mud, in which, with my own eyes, I have seen teams of mules buried.

When the soldiers returned to their camps, many found their huts flooded. The exhausted army coped as best they could in their soggy en-

vironment. Two days later, on January 25, President Lincoln relieved Burnside and named Joseph Hooker to the command. Never again would a major military maneuver or campaign occur during the winter months in Virginia.

The storm that stopped the Union army was a strong, winter coastal storm. The storm generated heavy rain along the coast and piedmont while heavy snow fell far inland. The total rainfall in Washington was 3.20 inches and the lowest barometric pressure was 29.75 inches. Gale-force winds from the northeast accompanied the storm on January 22 and 23. In Tennessee and Ohio, heavy snow fell and slowed military movements in those states. East of the Blue Ridge Mountains, temperatures remained fairly constant in the upper 30's and the precipitation fell as rain.

Light rain and fog continued for three days following the storm. The rain turned to wet snow on January 28 and continued through January 29. The total snowfall accumulation was insignificant, but the liquid total for the two days was an additional 0.88 inches, concluding a very soggy period in American military history.

The Great Snowball Battle of Rappahannock Academy, February 25, 1863

Two back-to-back snowstorms in February of 1863 provided the ammunition for a friendly snowball battle amongst rival divisions of Confederate troops near Fredericksburg, Virginia. On February 19, eight inches of snow fell on the region. Two days later, nine inches of snow fell. On February 25, sunny skies and mild temperatures softened the deep snow cover, providing ideal conditions for making snowballs.

During this time, the Confederate Army was camped near Fredericksburg. Some of the Divisions of the army had been reorganized, which had created friendly rivalries between the Confederate brigades and regiments. This helped spark a huge snowball battle near Rappahannock Academy in which approximately 10,000 Confederate soldiers participated. One soldier who participated in the snowball battle described it as "one of the most memorable combats of the war."

The battle started on the morning of February 25, 1863, when General Hoke's North Carolina soldiers marched towards Colonel Stiles' camp of Georgians, with the intent of capturing the camp using only snowballs. The attacking force, composed of infantry, cavalry and skirmishers, moved in swiftly. Battle lines formed and the fight began with "severe pelting" of snowballs. Reinforcements arrived from all sides to assist the brigade under attack. Even the employees of the commissary joined the snowball battle. Soon, the attacking soldiers were pushed back.

Hoke's beaten soldiers retreated back to their camp. Colonel Stiles then held a Council of War on how best to attack the retreating force. He decided to organize his men and march directly into their camp, with snowballs in hand. When Stile's forces finally arrived in Hoke's camp, they were quite surprised to find that their adversaries had rallied and filled their haversacks to the top with snowballs. This allowed Hoke's soldiers to provide an endless barrage of snowballs "without the need to reload." The attacking force was quickly overwhelmed and many of their soldiers were captured and "whitewashed" with snow. The snowball battle came to an end and both brigades settled back into their respective camps. The captured prisoners were quickly paroled and returned to their camp, to much heckling from fellow soldiers. It was noted that General Stonewall Jackson had witnessed the snowball battle. One soldier remarked that he had wished

Over ten thousand Confederate soldiers engaged in a spirited snowball battle near Fredericksburg, Virginia on February 25, 1863. Twelve inches of snow cover combined with mild temperatures provided ideal conditions for the massive snowball fight. Combatants employed real battle tactics in the snowball fight, including forming battle lines, charging, skirmishing, use of cavalry, and capturing prisoners. *Library of Congress*

Jackson and staff had joined the fight so he could have thrown a snowball at "the old faded uniforms."

The weather turned mild and rainy in the following days. Other snowball battles were documented during the Civil War – including a snowball fight at Dalton, Georgia – but The Snowball Battle of Rappahannock Academy was unique in size, strategy and ample snow cover. The depth of the snow cover on the day of the battle was documented in a soldier's diary to be 12 inches.

A snowy Capitol scene, February 10, 1926. The snowfall in Washington was 9.3 inches with a high temperature of 31°F and a low temperature of 25°F. *Library of Congress*

CHAPTER TWO

WINTER STORMS
AND BLIZZARDS

ituated between the Blue Ridge Mountains to the west, and the Chesapeake Bay and Atlantic Ocean to the east, the Washington metropolitan area is located in a classic "meteorological battle zone" in the winter. The battle pits dry, Arctic air which plunges south out of Canada against relatively warm, moist air that streams in from the Atlantic Ocean and the Gulf of Mexico. This often results in forecasts for the area which include the phrase "wintry mix," referring to a combination of snow, sleet, and freezing rain. In fact, it's not unusual for a winter forecast to call for four to eight inches of snow in places like Leesburg, Virginia and Damascus, Maryland, while 30 miles to the southeast in Washington the forecast calls for "snow changing to rain" with an accumulation of only one to three inches. When it comes to winter storms, a temperature fluctuation of just a couple degrees can turn a sluggish commute on the Beltway into an icy gridlock.

Wintry Precipitation Types

To fully understand why one day you may have snow falling with a surface temperature of 40°F, while on another day freezing rain may be falling with a temperature of only 25°F, you have to look at the bigger picture. Knowing how the temperature of the air changes above the ground is crucial in determining precipitation type. When

temperatures throughout the atmosphere are at or below freezing (32°F), snow will fall. In contrast, when temperatures throughout the atmosphere are above the freezing mark, rain will fall. However, the scenario is often more complicated than this. For example, when a storm moves to the northwest of Washington – let's say just west of the Appalachians – warm air will usually be swept in by winds from the south and southeast aloft. The inflow of warm air from the south erodes the deep layer of cold air, producing what is called a *temperature inversion*. In this case, the inversion refers to a warm layer of air somewhere between the surface and an altitude of 5,000 feet. The precipitation begins falling as snow in the cold layer above the inversion, but then the snow melts to rain as it encounters the warm layer. Depending on the depth of the subfreezing air below the inversion, the precipitation may refreeze into ice pellets (*sleet*) or fall as rain that freezes on contact (*freezing rain*) with objects near the ground that are at or below freezing.

Freezing rain and sleet occur about as often as snow in the Washington area during the winter. In fact, aside from central Pennsylvania and some of the deeper valleys in upstate New York, western parts of our region see more icy weather than just about anywhere else in the country. Parts of the Shenandoah Valley, and areas just east of the Blue Ridge in Maryland and Virginia, average between 30 and 40 hours of freezing rain each winter.

Winter Precipitation

Cold Air

RAIN   FREEZING RAIN   SLEET   SNOW

Snow Melts . . .   Warm Air

Snow Never Melts

Cold Air

Rain Freezes on Impact   Rain Refreezes into Sleet

Slip-Sliding Away

On January 14-15, 1999, a crippling ice storm struck the nation's capital and its surrounding suburbs. While temperatures aloft were too warm for snow, a dry, arctic air mass was in place near the ground. As the rain fell into the dry, cold air, it initially evaporated, further cooling and reinforcing the arctic air. As rain fell through the night, it increased in intensity and froze on everything. Trees and power lines were no match for the 1 inch thick layer of ice that coated them. Hundreds of thousands of people in the Washington area lost power, thousands of trees were toppled, and many roads were impassable for days! Yet, just south of the District, in places like Waldorf and Upper Marlboro, the freezing rain event did not occur; temperatures were a few degrees warmer and only rain fell.

The "Alberta Clipper"

Often producing what may be termed "nuisance snow," the "Alberta Clippers" are fast-moving, low-pressure systems which are enhanced on the lee side of the Canadian Rockies in southwestern Canada (Alberta). They usually track southeasterly into the Northern Plains, through the Upper Midwest, and then zip across the Northeast or Middle Atlantic Region. Due to their quick movement and great distance from a moisture source (like the Gulf of Mexico), clippers usually result in only light snow, followed by a blast of colder air. However, there are exceptions to every rule. Just a few days after the great "Blizzard of 1996" struck the Middle Atlantic region, a rather vigorous Alberta Clipper moved through the Washington area. Accompanied by strong winds in the upper atmosphere which helped to lift the air, this Clipper added to the misery by dumping up to five inches of new snow on roads that were still being cleared of nearly two feet of snow that had just fallen with the blizzard.

Nor'easters: Winter's White Hurricanes

One of the first weather watchers to gain a true glimpse into the nature of nor'easters was Benjamin Franklin. In 1743, while staying up late one night to watch a lunar eclipse in Philadelphia, the weather turned stormy and prevented him from viewing it. Later, he learned that his brother in Boston had seen the eclipse as scheduled, but had noted that he was hit by the same storm later that night. Franklin was puzzled as to why the storm would move against its prevailing winds, which blew from the northeast. After thinking about it, he concluded that the storm's winds must circulate counterclockwise. This explained why the prevailing winds in Philadelphia blew from the northeast, while the storm itself moved toward the northeast.

Today, we know a lot more about the type of storm that spoiled Franklin's view of the eclipse. In fact, Washington's biggest and most famous winter storms are those that form along the coast. These storms are called *northeasters,* or as they are more commonly known, *nor'easters.* Their name is derived from the strong northeast winds that are generated ahead of the storm as air circulates in a counterclockwise direction around the storm center. Many of the strong nor'easters that bring significant snow to Washington form in the northern Gulf of Mexico or along the southeast U.S. coast and then move up the eastern

seaboard; some that originate in the Gulf of Mexico track into the Ohio Valley before redeveloping east of the Carolinas.

During winter, a fierce coastal storm can produce snowfall rates up to 4 inches per hour, with thunder and lightning, while 30-40 mph winds can pile snow in five to ten foot drifts. Winds are usually much stronger near the coast, often exceeding 60 mph. Nor'easters are also notorious for creating relentless, pounding waves that can demolish oceanfront homes and wash away miles of beach. The "Ash Wednesday Storm" of 1962, arguably one of the worst nor'easters of the 20th Century, assaulted the East Coast for five days in early March! While 20-25 foot waves and 60 mph winds pummeled places like Ocean City, Maryland and Atlantic City, New Jersey, a blinding snowstorm raged in the mountains. Big

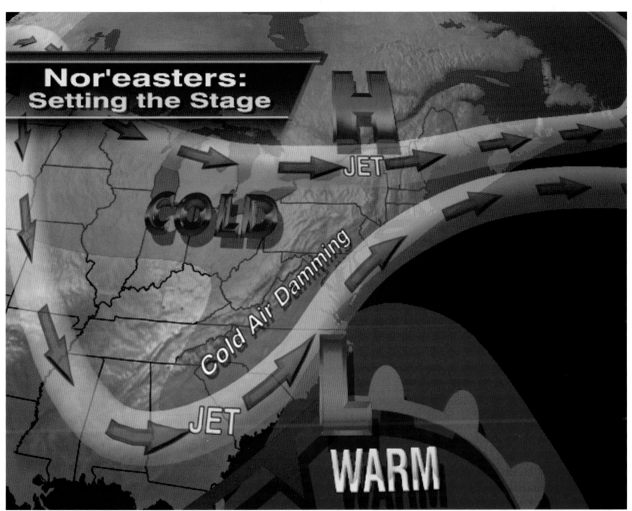

Meadows, southeast of Luray, Virginia, was buried by 42 inches of snow, a state record that still stands today.

Nor'easters: Setting the Stage with Cold High Pressure to the North

The classic setup for a nor'easter begins as a cold dome of high pressure builds over New England and/or Quebec, Canada. The importance of the high is two-fold. First, it serves to impede the northward movement of the developing storm. A slow-moving storm results in a long-duration precipitation event and produces strong winds blowing in the same direction over several tide cycles. This creates large, destructive waves along the coast. Second, the high acts as a conduit for the cold air. As the cold air flows around high pressure in a clockwise direction from New England toward the Appalachian Mountains, it is too dense to make it over the high terrain. As a result, it takes the path of "least resistance," which results in the cold air funneling southward through Pennsylvania, into Maryland, Virginia, and the Carolinas. This is called *cold air damming.*

Nor'easters: A Favorable Jet Stream

Once a supply of cold air is established, a sequence of events begins to take place high in the atmosphere to spark *cyclogenesis* (storm formation). A favorable jet stream pattern for big nor'easters to hit Washington is called a split-

---

THE JET STREAM

The *Jet Stream* is a relatively narrow band of high-speed winds that circle the earth in each hemisphere. Jet Streams are produced by large differences in temperature between the equator and the poles. The greater the temperature contrast, the greater the pressure difference (pressure gradient) between the warm and cold air. The large pressure gradient creates strong winds aloft, often in excess of 100 mph. As an analogy, consider a bathtub partially filled with water. As you slosh the water back and forth, you create a pressure gradient from one end of the tub to the other. Water pressure is higher where the water is deeper, and lower at the shallow end. Consequently, water flows from the deep end to the shallow end to create uniform water pressure. Similar to the water, the wind blows from higher pressure to lower pressure to restore balance in the atmosphere. (Note that while the *pressure gradient force* initiates the wind, there are actually several other forces that influence the motion of air.) Over the United States, the Jet Stream is usually found between 25,000 and 35,000 feet of altitude during winter.

---

flow pattern. This is when the jet stream splits after reaching the west coast of the U.S. The northern branch travels across Canada toward New England. It helps maintain a dome of cold air over the Northeast and Middle Atlantic region. The southern branch crosses the Rockies, dives southward toward the Gulf Coast, and then makes a sharp turn to the northeast across the Carolinas and then off the New England coast. As a disturbance moves through the southern branch of the jet stream toward the Gulf Coast, the winds become stronger and air is swept away aloft. This is referred to as *divergence.* This forces warm, moist air to rise up from the surface to replace it. You have probably witnessed this at home while sitting in front of a roaring fire in the fireplace. As the wind blows across the top of the chimney outside, air is forced to rise up through the chimney.

Nor'easters: A Storm is Born!

In nature, divergence aloft can cause the atmospheric pressure to fall at the surface, depending on other factors. Once the pressure begins to fall, air spirals in to the center of the low-pressure area, and the storm intensifies. The Gulf Coast and the offshore waters of North and South Carolina are a prime breeding ground for nor'easters. The storms usually form along coastal fronts that develop due to the large temperature contrast between the cold land and the warm water just offshore. January water temperatures about 50 miles offshore are often a balmy 68°F to 78°F in the Gulf of Mexico and in the Gulf Stream east of the Carolinas; meanwhile, land temperatures can be in the 20's or 30's! A coastal storm fueled by very warm, moist air over the Gulf Stream, coupled with a powerful upper air disturbance, can result in explosive development, leading to a very deep, rapidly intensifying low-pressure system called a *Bomb*.

Slight Storm Track Changes Can Have Huge Implications!

As the storm moves up the coast, its counterclockwise circulation brings moisture-laden air streaming in from the Atlantic Ocean. Since the warm air is lighter and more buoyant than the cold, dense air over land, the warmer air is forced to rise. As it rises, the air cools and its moisture condenses and clouds form. In fact, a large cloud

shield covering most of the eastern United States is often observed. Within this cloud shield, precipitation forms and eventually begins to fall.

Since the band of heaviest snow that accompanies a typical nor'easter is usually only 50-miles wide, predicting the storm's precise track and speed is crucial for making an accurate forecast. A storm that moves 50 miles farther east or west than expected, can spell the difference between an inch of rain and a foot of snow falling in a particular location! A track that "hugs" the coast brings more warm air inland than a storm that stays offshore. This influx of warm air often changes the snow to rain on Maryland's Eastern Shore, while a wintry potpourri of snow, sleet, and rain falls in an area from Washington east to the Chesapeake Bay. In addition, the band of

Aerial view of the snow-covered Pentagon, December 13, 1960. On the previous day, 8.5 inches of snow fell on Washington with a high temperature of 26°F and a low temperature of 14°F. *Copyright Washington Post; Reprinted by permission of the D.C. Public Library*

heavy snow often shifts farther west toward the northern and western suburbs of Washington and the mountains where the air remains at or below freezing through a large portion of the lower atmosphere. This was the case during the March "Super Storm" of 1993. While heavy rain and high winds socked the Eastern Shore, one and one-half feet of snow fell near Frederick, Maryland. On the other hand, a storm track farther offshore results in less warm air and less moisture moving inland. This

may result in blizzard conditions for Washington and points east, while snow amounts diminish rapidly to the north and west of Washington. This is what happened during the "Presidents' Day Storm" of 1979. One and one-half feet of snow fell in Washington, with a little over two feet in parts of Delaware and the lower Maryland Eastern Shore. Meanwhile, only light snow and flurries were reported in many areas west of the Blue Ridge.

Dreaming of a White Christmas? Better Head North and West!

On average, only about 18 inches of snow falls in the Nation's Capital in a typical winter season. Winter snowfall averages range from about ten inches at Rehoboth Beach, Delaware, and Ocean City, Maryland, to over 40 inches in the higher elevations on Skyline Drive in Virginia. A typical winter will yield close to two feet of snow for much of the northern and western suburbs of Washington, including Manassas, Sterling, and

Gaithersburg; about 20 inches in Bethesda and Arlington, and only about 15 inches for places near the Chesapeake Bay, such as St. Mary's City and Deale.

Dreams of a White Christmas in Washington are often dashed. The chance of seeing one is only about 10% in a given year. However, there's an old saying that perhaps best characterizes the "Jekyll and Hyde" nature of Washington winters: "Climate is what you expect; weather is what you get." During the winter of 1995-1996, Washington was buried by 46 inches of snow, yet the following two winters the city received a total of less than 7 inches of snow in Washington.

Occasionally, seasonal snowfall totals are inflated by one or two big storms that can deliver a winter's worth of snow in less than 24 hours. From 1899 to 1999, there were a dozen storms that dropped a foot or more of snow in the Washington area. That equates to only about one such event every eight years. So, snow lovers in Washington must either patiently wait for the "big storm" or be willing to head north or west to the mountains for more of the white stuff!

WINTER STORM EVENTS

The Blizzard of 1888:
The White Hurricane

The Blizzard of 1888 began in Washington early on March 11 with heavy rain that filled gutters and flooded streets. The rain changed to snow about 3:00 p.m. and continued into the night. As the temperature fell and the wind increased, telegraph and electric wires began to come down throughout the city. By midnight, the city was completely blacked out, with the exception of a few gas streetlights that flickered weakly in the furious storm.

The storm continued through the night. By the morning of March 12, the snow had ended. The snow depths varied from a few inches within the city to over 10 inches in areas well to the north and west. In the Northeast, the storm was particularly severe, with New York City receiving 21 inches and Albany, New York receiving 46 inches.

High winds continued for days after the storm. On the morning of March 13, the wind reached a maximum speed of 48 mph in Washington. An unprecedented number of telegraph, electric, and police wires had been blown down throughout the city. The entire area was cut off from all communications. Public officials were outraged that weather could cause such a widespread communication outage.

A reporter from the Evening Star colorfully summed up the lack of communications in Washington as follows: "Washington experiences the peculiar sensation of being snowbound. In certain respects the city is isolated by the effects of snow and wind. It is also cut off from the telegraphic communications with the outside world, and the daily picture of the doings of distant states and empires."

After the storm, the winds blew from the north for such a long duration that much of the water was blown out of the Potomac River. The bottom of the river was exposed from the shoreline out to the channel and dust clouds blew down the dried-out riverbed. Boats and steamers were grounded high and dry. "Old salts" were overheard to report that they had never seen the Potomac in such a state. Near Washington, the Potomac River was more than five feet below normal low-tide levels. In Baltimore, the Chesapeake Bay was twelve feet below normal low-tide levels and the bottom of the harbor was exposed.

The storm had originated in the Southeast and had tracked north along the East Coast. It stalled off the coast of New York and even looped back before moving away. The Northeast was caught in a prolonged period of precipitation while the Middle Atlantic had a quick, heavy dose of rain and snow, then a prolonged period of wind. In Washington, the snow fell for less than nine hours on March 11. Cold, windy weather followed

---

Washington City, Sunday, March 11, 1888—7 A. M.

Indications for 24 hours, commencing at 3 P. M., Sunday, March 11, 1888.

Fresh to brisk easterly winds, with rain, will prevail to-night, followed on Monday by colder brisk westerly winds and fair weather throughout the Atlantic states; colder fresh westerly winds, with fair weather, over the lake regions, the Ohio and Mississippi valleys; diminishing northerly winds, with slightly colder, fair weather, in the Gulf states; light to fresh variable winds, with higher temperature, in Kansas, Nebraska, and Colorado.

SIGNALS.—Cautionary southeast signals are displayed on the Atlantic coast from Norfolk section to Wood's Holl section.

RIVERS.—The rivers will rise slightly.

The forecast issued for Washington City, March 11, 1888. Although the forecast called for rain, a blizzard ensued. Reporters later pressed forecasters for an explanation about the surprise blizzard; however, Washington's chief forecaster was unavailable for comment. *NOAA Library*

Capitol snow scene after the Blizzard of 1888. A station attendant has cleared an area next to the tracks to allow passengers to board Washington's horse-drawn streetcars. The blizzard produced strong winds that knocked down thousands of telegraph wires in the area, cutting off all communication into and out of the city. Snowfall totals ranged from 2 to 10 inches, with accumulations heaviest to the north and west of D.C. *Library of Congress*

the blizzard. On March 12, Washington's high temperature was only 19°F and the low temperature was 11°F.

The Blizzard of 1899: Washington's "Snow King"

The Blizzard of 1899, dubbed the "Snow King" by the local media, capped off a week of extremely cold, snowy weather. The harsh weather began on February 5, when a fast moving low-pressure system from Louisiana dumped between 5 and 6 inches of snow on the area. A second low moved in from the south and dumped an additional 2 to 3 inches of snow on February 6 – bringing the snow cover in Washington up to 8 inches. A third storm brought snow back to the area on February 7. That storm added almost 5 inches to the snowpack. With temperatures that were generally in the twenties during the three storms, there was a very brief period of freezing rain at the end of the third event.

Soon after the third storm had passed, record

A snowdrift on H Street, across from the Government Printing Office after the Blizzard of 1899. Snow fell in Washington for 51 hours, from February 11 to February 14, accumulating 20 inches. High winds during the blizzard produced ten-foot snowdrifts. *Washingtoniana Division, D.C. Public Library*

Deep snow in front of the Music Academy at Ninth and D Streets after the Blizzard of 1899.
A series of snowstorms dropped 34.9 inches of snow in Washington and were accompanied by
very cold temperatures and strong winds. Washington's all-time record low of –15°F occurred the
day before the blizzard on February 11, 1899. *Washingtoniana Division, D.C. Public Library.*

cold moved in from Canada. On the morning of February 11, Washington set its record low of -15°F. Fredericksburg and Quantico recorded -21°F. A few unofficial readings in the Washington area were as low as -25°F.

The big snow event, however, did not begin until the morning of February 12 when a storm began to track up the East Coast from Florida. The storm developed rapidly and passed just east of Norfolk on the morning of February 13. The storm was at its peak on the afternoon of February 13 when wind speeds reached 35 mph, with gusts of 48 mph. Temperatures remained in the single digits throughout the storm. By the evening of February 13, the snow had ended.

Snowfall
February 5-14, 1899

At the conclusion of the storm, 34.2 inches of snow lay on the ground in Washington. Of that amount, approximately 20 inches of snow fell during the blizzard of February 12 and 13. The liquid equivalent of all of the snow events was 3.69 inches.

Over 40 inches of snow was on the ground in areas south and east of Washington. Drifting was generally around 4 to 6 feet; however, some drifts were as high as 15 feet. In areas of upper Montgomery County, railroad cuts filled with drifting snow to a depth of 20 feet.

Generally, 1 to 3 feet of snow fell from eastern North Carolina to southeast Massachusetts. A foot of snow fell in Columbia, South Carolina and up to 4 inches of snow accumulated in north Florida. All-time low temperature records include the following: -39°F in Pennsylvania and Ohio; -9°F in Atlanta, Georgia; -8°F in Dallas, Texas; -5°F in Montgomery, Alabama; and -2°F in Talla-hassee, Florida (the only time Florida has ever been below 0°F).

The Taft Inaugural Snowstorm of March 4, 1909

On March 3, 1909, Washington swelled with dignitaries and visitors, eagerly anticipating the next day's Inauguration of the newly-elected President, William Howard Taft. It appeared that weather would not be a factor. The headline in the March 3 Evening Star proclaimed "fine weather tomorrow" and the Weather Bureau promised, "ideal marching weather."

During the afternoon of March 3, heavy rain pelted the area, accompanied at times by lightning and thunder. Much to everyone's surprise, the rain changed to snow during the evening. The forecast for the Inauguration, however, continued to call for fair weather. The wind direction in Washington had already changed from east to west, indicating that the storm center had passed to the north and away from the area. The head of the Washington Weather Bureau personally called Mr. Taft at midnight and stated that the snow would soon be over. He reiterated his promise that the weather would not interfere with any of his inaugural activities.

To the chagrin of forecasters, the snowstorm picked up in intensity during the predawn hours and by morning a near whiteout was in progress, accompanied by a howling northwest wind. The storm center had rapidly intensified over southern New Jersey, giving Washington an unusually heavy wraparound snow event. Indeed, the explanation later given by the Washington Weather Bureau regarding the incorrect forecast was that the snow had "flared back" from southern Pennsylvania after the storm center had already passed well to the north of the area.

An army of six thousand shovelers worked tirelessly trying to clear the parade route as the storm raged on. The stands along Pennsylvania Avenue were virtually deserted. One Congress-man described the scene as "the worst weather on the face of the earth."

Due to the weather, President Taft moved the planned outside "oath of office" ceremony to inside the Senate chamber. Later he joked, "I always knew it would be a cold day in hell when I became President." The snow stopped within minutes of the noontime indoor swearing-in event. For a while, there was question whether there would be any parade at all, but despite the whistling winds, 20,000 marchers braved the elements and trudged down the snowy parade route in the afternoon. Hundreds of trains destined for the inaugural festivities were delayed due to the weather and many spectators did not arrive until the parade was well underway.

The Weather Bureau in Washington reported that 10 inches of snow fell during the storm. Snow amounts were generally heavier to the north and east, with Laurel, Maryland receiving 13.5 inches, Annapolis, Maryland receiving 14 inches, and

President Roosevelt and President-elect Taft ride inside a horse-drawn carriage en route to Taft's Inauguration, March 4, 1909. Almost a foot of wet snow fell in Washington prior to the Inauguration.
Library of Congress

Towson, Maryland receiving 16.8 inches.

The Knickerbocker Snowstorm of January 27-28, 1922

Washington's largest snowstorm on record began during the evening of January 27, 1922. By the morning of January 28, the snow total had reached 18 inches. By mid-afternoon, the accumulation reached a depth of 25 inches. The snow did not stop until the morning of January 29, with an official snow depth of 28 inches, a single storm snowfall record that still stands today. A snow depth of 33 inches was

measured in Rock Creek Park, three miles to the north of Washington's official weather station. Temperatures were in the low-to-mid-20's during most of the storm. The liquid total of the snowfall was 3.02 inches.

The weight of the record-breaking snow

Crandall's Knickerbocker Theatre on the morning after its roof collapsed under the weight of a 28-inch snowfall, January 29, 1922. The roof collapsed during a show on the evening of January 28. This photograph was taken from the police line and shows ambulances waiting to take away the injured. The death toll was 98, with 133 people injured. *Washingtoniana Division, D.C. Public Library*

THE COLLAPSE OF THE KNICKERBOCKER THEATRE'S ROOF
JANUARY 28, 1922

On the evening of January 28, 1922 several hundred people fought their way through a massive snowstorm to see the show at the Knickerbocker Theatre, Washington's largest and most modern moving picture theater of the time. When the show began that evening, the greatest snowstorm in Washington's history was winding down. It had already dumped over two feet of heavy, wet snow on the city and many flat-roofed buildings, like that of the Knickerbocker Theatre, were tremendously burdened by the weight.

Shortly before 9:00 p.m., the Knickerbocker Theatre's orchestra was playing for intermission. The lights had dimmed and the people were returning to their seats. Suddenly, a loud hissing noise filled the room. The ceiling, weighed down from the snow, had begun to split apart down the middle. The few people who had noticed the splitting ceiling dove under their seats or ran for the door. Within seconds, the entire roof started to fall towards the crowd. As the roof came down, it

Digging through the rubble inside the Knickerbocker Theatre, January 29, 1922. *Library of Congress*

collapsed the theater's cement balcony and pulled down portions of the theater's brick wall. Concrete, bricks and metal crashed to the ground, burying dozens of people.

George Brodie had entered the theater moments before the roof collapsed and gave the following account: "I grabbed for my hat and coat, and the next minute found myself flat on my face with something weighty on top. I lay still for about five minutes when I noticed at the side of me a girl with an arch or pillar resting upon her. I tried to pull it off but couldn't move it. Then I started working my way slowly in some direction – I think the middle – and with four other fellows we saw a hole with a light shining through. The next thing I know I was on the street, but I don't know how I got there. I stayed around for a while and helped several others, who were apparently uninjured, out of the place. It was a frightful sight within, nothing but moans, cries and darkness."

The scene after the disaster was terrible. People ran through the ruins calling for missing loved ones. Shouts from rescue workers mixed with the cries of anguish from victims buried under the wreckage. Lanterns and shadows could be seen darting about through the heavily falling snow. Great masses of twisted steel, splintered timber and crumbled masonry covered the floor of the theater. One reporter wrote that no description of the scene could convey the awfulness of what he had witnessed that night. Another reporter, with recent memories of the devastation of World War I in mind, wrote, "Stark and grim as any ruin in the war-swept area of France or Belgium stood the walls of the Knickerbocker theater."

The chaotic rescue effort became better organized when the police and firemen arrived at the scene. Police lines were drawn and heavy equipment was called in. By 12:00 a.m., 200 police, soldiers and firemen were working feverishly, digging through the wreckage. By 2:30 a.m., over 600 rescue workers were on the scene. Residents in the vicinity of the theater supplied hot food and coffee to the rescuers.

The workers had to dig through two layers of debris to rescue the injured. First they had to remove the plaster and steel of the roof to reach the injured that had been seated in the balcony. Large saws were used to cut through the roof's heavy wire screen that had once held the ceiling's plaster. After the roof had been removed, the workers had to chisel through the cement structure of the balcony to rescue those who had been seated on the first floor. The rescue effort was not completed until the following afternoon.

The toll for the disaster was 98 dead and 133 injured. Every hospital in the area was filled with the injured. Many stores and houses served as short-term first-aid stations. Hotels opened their doors to the injured as well as the rescuers. The disaster ranks as one of the worst in Washington's history. The snowstorm still ranks as Washington's largest single snowfall.

Digging out and commuting during the record-breaking Knickerbocker Snowstorm, January 28, 1922. A slow moving storm system dumped 28 inches of snow on Washington. Temperatures hovered in the 20's for the majority of the storm, but slowly rose to 31°F as the snow ended. *NOAA Library*

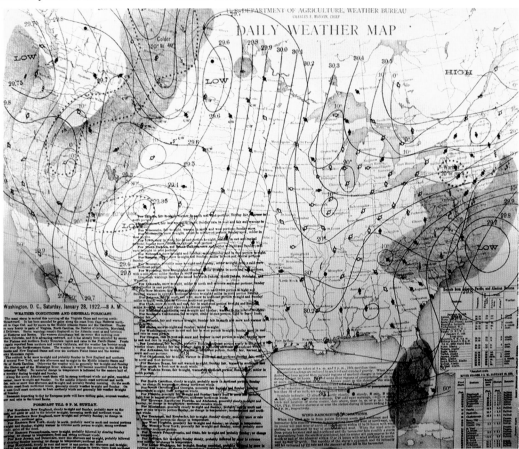

The surface weather map for January 28, 1922 shows the Knickerbocker Snowstorm east of Cape Hatteras, North Carolina. The storm moved east-northeast out to sea after dumping a record snowfall on Washington. The shaded area on the map represents precipitation. *NOAA Library*

Members of Washington's snow removal force shovel snow into a wagon, December 1932. Over 5000 men, unemployed due to the depression, helped clear the city streets. Twelve inches of snow fell in D.C. on December 17, and this remains one of Washington's largest December snowfalls. *Library of Congress*

collapsed the roof of the Knickerbocker Theatre. The roof of the theater fell on scores of moviegoers, killing 98 and injuring 133. The disaster ranks as one of the worst in Washington's history.

The storm responsible for the record snowfall formed east of South Carolina on the morning of January 27 and moved slowly north to a position well east of Cape Hatteras on the morning of January 28. It then drifted slowly east-

northeast out to sea. A stationary high-pressure system north of New York State ensured that temperatures remained cold throughout the event.

The Snowstorm of December 17, 1932

The snowstorm of December 17, 1932 was a classic Washington snowstorm. The storm originated over north Florida and moved up the East Coast as a high-pressure system remained anchored over New York and New England. The heaviest snow fell in a band that was centered across the D.C. area.

The snow began early on December 17 and fell at a rate of 1 to 2 inches per hour until afternoon. Twelve inches of snow accumulated

The Capitol under a mantle of snow, January 24, 1935. Initially, the precipitation started off as a cold rain on January 22, then transitioned to snow on January 23. The storm dropped 11.3 inches of snow in Washington as temperatures dropped into the upper 20's.
Washingtoniana Division, D.C. Public Library

throughout the area. It was a rather cold storm with temperatures hovering around 18°F during the height of the snowfall. Temperatures rapidly warmed after the storm, and the snow was gone by Christmas.

Despite the heavy snowfall, Washington's roads were quickly cleared. An army of unemployed men, jobless due to the depression, took to the streets with snow shovel in hand. Over 5000 men worked on the roads for days, and were paid 35 cents an hour for their labor. The District's streets were cleared so thoroughly that the Weather Bureau complimented the city on the excellent condition of the roadways.

Horse and sleigh in Rock Creek Park, January 24, 1935. D.C. Police closed twenty blocks for sledding and sleigh riding after the storm. Horse and sleigh rides around Washington were a recreational and fashionable form of transportation during the 19th and early 20th centuries. *Washingtoniana Division, D.C. Public Library*

The Snowstorm of January 23, 1935

The snowstorm of January 23, 1935 came at the end of a long, mild spell. The day before the storm, the temperature reached 50°F before a cold front moved slowly through the area with rain. As the front moved

Street scene looking towards Lafayette Park, February 7, 1936. A snowstorm dropped 14.4 inches of snow in Washington with temperatures that hovered in the upper teens. The heaviest snow fell to the southeast of D.C., with 18 inches of snow recorded in southern Maryland. *Washingtoniana Division, D.C. Public Library*

Maryland, with 14 to 18 inches of snow recorded. In southern Maryland and on the Eastern Shore, the precipitation fell as freezing rain, resulting in a severe ice storm.

The storm was followed by a long cold wave that lasted until the end of January. The coldest reading occurred on January 28, when the temperature plunged to a low of -2°F. Washington would not experience another sub-zero temperature for 48 years – not until January 1982.

The Blizzard of February 7, 1936

The Blizzard of 1936 commenced when a low-pressure system east of Florida on the morning of February 6 moved to position about 60 miles east of Cape Hatteras, North Carolina on the morning of February 7. The first flakes fell in Washington just after midnight on February 7, and by late morning, over a foot of snow had fallen. Light snow continued throughout the afternoon, ending during the evening. The total accumulation was 14.4 inches, with a liquid content of 1.01 inches. Very cold weather accompanied the storm as temperatures fell to 16°F during the height of the snowfall.

Washington was on the northwest edge of the heavy snow band, with up to 18 inches of snow falling in southern Maryland. Only 3 to 6 inches of snow fell in the far northern and western suburbs. Norfolk recorded 9 inches of snow, the city's heaviest snowfall in over 40 years.

Blowing snow grounded all air traffic at Washington Airport on February 7. The airport remained closed on February 8 while the snow

east, a low-pressure wave formed along the southern end of the front near Alabama. As the low moved into western North Carolina, a second low pressure system developed near Nags Head. With high-pressure over northern New England, the pressure gradient tightened and a moist easterly fetch developed off the Atlantic.

Rain continued into the evening of January 22, but temperatures began to fall. Temperatures fell below freezing shortly after midnight, and sleet and freezing rain began to fall. The precipitation changed to all snow on the morning of January 23, with the temperature falling to 28°F. Snow fell moderately to heavily all day on January 23, approaching two inches per hour. The snow accumulation was 11.3 inches, with a liquid total of 1.57 inches – rain, ice and snow.

The heaviest snow totals fell over northern

Snowdrifts along a road near Rockville, Maryland, February 1940. Cold weather and a series of snowstorms produced heavy snow cover throughout the Washington area during January and February 1940. The heaviest snowfall occurred on January 24 when 9.5 inches of snow fell in Washington and 21.3 inches fell in Richmond, Virginia.

Library of Congress

was cleared from the runways. Mail service was also curtailed on February 8.

The Snowstorm of January 23-24, 1940

During late January 1940, Washington was on the northern fringe of one of the greatest snowstorms ever to hit the South. Snowfall totals included 21.3 inches in Richmond, 22 inches in Danville and 8.3 inches in Atlanta, all of which set 24-hour snowfall records in those cities. Officially, 9.5 inches of snow fell at Washington's weather recording station (at that time located near 24th and M Street); however, over a foot of snow fell in southeast Washington.

Snow began during the evening of January 23 and was over by mid-morning of January 24. Most of the snow fell between 2:00 a.m. and 5:00 a.m., when snow came down at a rate of 2 inches per hour.

The storm system tracked up the East Coast to near Cape Hatteras, North Carolina, and then swerved out to sea. As a result of the easterly storm track, northeastern cities were spared a major accumulation from the storm.

There was an extreme gradient of snowfall within the Washington, D.C. area. In Frederick and Upper Montgomery Counties, only a few inches of snow fell. However, just east of Andrews Air Force Base, 24 inches of snow was measured. Generally, 20 to 25 inches of snow fell in southern Maryland, as well as central and eastern Virginia. High winds caused 10-foot drifts, crippling the area for days. A cold wave after the storm lasted until the end of the month, with a high of only 20°F on January 26.

The Spring Snowstorm of March 29, 1942

During the evening of March 28, the temperatures in D.C. were in the mid-40s and light rain was falling. Temperatures fell during the evening and by midnight the rain had turned to snow. Most of the accumulating snow fell during the predawn hours of March 29, with temperatures hovering at 32°F. Light snow continued all day on March 29, but did not add much to the snow cover. On March 30, temperatures rebounded to 54°F and the snow rapidly melted.

Snowfall tallies varied considerably across the Washington area. At Quantico, Virginia, only a trace of snow fell. In Washington, 11.5 inches of snow fell. The snow depths picked up markedly to the north. College Park received 15 inches and Takoma Park and Silver Spring received over 18 inches. Farther north, Laurel received 20.3 inches and downtown Baltimore received 22 inches. An impressive 32 inches of snow fell at Westminster, Maryland and between 35 to 40 inches of snow accumulated at State College, Pennsylvania – their greatest storm of all time. Almost no accumulating snow fell at Philadelphia and New York City.

Car on Randolph Street near 14th Street, March 29, 1942. Heavy, wet snow accumulated 11.5 inches while temperatures hovered near the freezing point. *Library of Congress*

A snowball fight in the park, March 29, 1942. A wet snow blanketed the Washington area with 11.5 inches in town and up to 18 inches of snow in the northern and western suburbs. The high temperature was 35°F and the low temperature was 32°F. *Washingtoniana Division, D.C. Public Library*

Struggling to make deliveries in the snow, March 29, 1942. Despite marginally cold temperatures and a high March sun angle, the snow fell hard enough to leave substantial accumulations on road surfaces. *Library of Congress*

Right: Military aide to the President, General Harry Vaughan, throws a snowball on the White House lawn, December 20, 1945. Six inches of snow fell on Washington. *Copyright Washington Post; Reprinted by permission of the D.C. Public Library*

Below: Sledding on Capitol Hill, January 24, 1948. Snowfall measured 9.5 inches in Washington, with a high temperature of 17°F and a low temperature of 10°F. *Washingtoniana Division, D.C. Public Library*

A massive snowball, Upper Marlboro, Maryland, November 7, 1953. An early season snowfall dropped 6.7 inches of snow on Washington while central Pennsylvania received as much as 2 feet of snow. *Copyright Washington Post; Reprinted by permission of the D.C. Public Library*

The Early Season Snowstorm of November 6, 1953

A tropical depression moved east from the central Gulf of Mexico and crossed Florida on November 5, accompanied by heavy rain and 30 to 40 mph winds. As it reached the Atlantic coast and passed out over the Gulf Stream, it rapidly intensified. At the same time, an unusually cold outbreak was enveloping the Midwest and Northeast. The storm turned sharply to the north and moisture from the storm was thrown back into a cold air mass, creating a band of snow that developed from North Carolina to New York.

In Washington, snow began to fall during the early morning hours of November 6. It remained light most of the day, but became heavy for awhile during the evening hours. With a tight pressure gradient set up between the strong ocean storm and the cold high-pressure system to the north, winds in Washington were unusually strong – averaging at times over 30 mph. As a result, there was extensive blowing and drifting of the snow.

Hundreds of stalled cars clogged the roadways throughout the area during the height of the storm. Drifts blocked many roads in Montgomery County. Colesville Road and Viers Mill Road were declared impassable and Old Kings Highway in Alexandria, Virginia was blocked by large drifts. Near Upper Marlboro, a 7-foot snowdrift was reported across Crain Highway. As a result of the blocked road, about 100 stranded travelers were forced to spend the night in the Prince George's County Courthouse.

Sledding in Upper Marlboro, Maryland, November 6, 1953. The snowfall at National Airport was 6.7 inches. *Copyright Washington Post; Reprinted by permission of the D.C. Public Library*

The snowfall at National Airport was 6.7 inches. This is one of Washington's biggest early season snowfalls, not far behind the Veterans' Day Snowstorm of November 11, 1987, which piled up 11 inches in D.C.

The Snowstorm of December 4, 1957

A weak wave of low-pressure moved east from Kentucky on December 3, 1957. When the storm emerged off of the Atlantic coast, it became energized by the warm Gulf Stream waters and developed very rapidly. Snow rapidly spread throughout the metropolitan area during the morning of December 4. The snowfall soon became heavy, with large flakes falling at an alarming rate.

The blinding snow continued into afternoon and finally ended during the evening, with 11.4 inches of snow measured at National Airport. That was the heaviest snowfall at National Airport since the Spring Snowstorm of March 1942. Up to 14 inches was reported in the suburbs. The snowfall took place with the temperature hovering at 32°F. This made for a wet snowfall, especially south of Washington where scattered power outages occurred.

The Snowstorm of February 15, 1958

The Snowstorm of February 15, 1958 began when a storm center developed over Texas on Valentine's Day. The storm dove rapidly southeast to a position near Mobile, Alabama on February 15, then it recurved to the northeast and began to intensify. The storm center passed over Columbia, South Carolina and deepened markedly as it moved northeast across eastern North Carolina and off the Virginia Capes. By midnight, on February 16, the low was just east of the Delmarva Peninsula, with a central pressure at an impressive level of 28.80 inches.

Between 14 to 18 inches of snow fell across the Washington area. The deep low-pressure system created damaging winds on the Delmarva Peninsula that disrupted power and blew down many trees. In Washington, winds were sustained at 30 to 40 mph for several days after the storm, which caused terrible drifting, in many instances at least six feet deep.

Snowfall rates during the storm were among the heaviest ever recorded at National Airport. Eleven inches of snow fell during a five-hour period on February 15, and over three inches of snow fell during a one-hour period.

Generally, a snowfall of 10 to 20 inches occurred from Alabama to Massachusetts. Up to 30 inches were reported in northeast Pennsylvania. The heaviest snow band was fifty miles wide and passed directly

Snow falls on the National Bank of Washington, December 4, 1957. One foot of snow fell on Washington, with up to 14 inches of snow in the suburbs. *Copyright Washington Post; Reprinted by permission of the D.C. Public Library*

Snow falls on the Old Post Office in Washington, December 4, 1957. Light snow developed in the morning of December 4, then quickly intensified, dropping 11.4 inches at National Airport. Temperatures hovered in the lower 30's. *Copyright Washington Post; Reprinted by permission of the D.C. Public Library*

A car owner in the 400 block of Tenth Street evaluates the job of digging out, February 15, 1958. The Snowstorm of 1958 produced very heavy snowfall rates, over 3 inches per hour. The snowfall at National Airport was 14.4 inches. *Copyright Washington Post; Reprinted by permission of the D.C. Public Library*

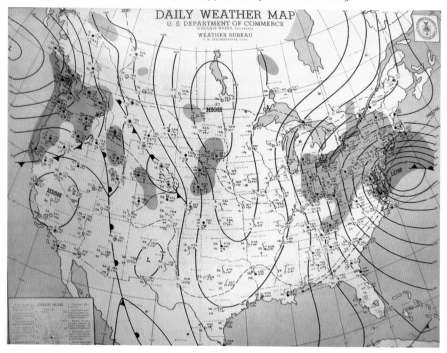

"Frosty Sahara on the Mall" after the Snowstorm of February 15, 1958. A very intense snowstorm dropped between 14 to 18 inches of snow across the Washington area. Cold temperatures and strong winds persisted for several days after the storm, causing significant drifting. *Copyright Washington Post; Reprinted by permission of the D.C. Public Library*

The surface weather map for February 16, 1958. The map shows the storm rapidly intensifying off the coast of Ocean City, Maryland. Solid lines are isobars, which correspond to lines of equal pressure. *NOAA Library*

Cars buried in snow near the Capitol, February 16, 1958. National Airport received 14.4 inches of snow with heavier amounts to the north. Boston, Massachusetts received 19.4 inches of snow and some areas of New England received in excess of 30 inches. Cold weather followed the storm and the ground in Washington remained snow-covered until February 23. *NOAA Library*

and by afternoon the downpour had changed to snow in the normally colder northern and western suburbs. During the late afternoon and evening hours, most areas switched back and forth between snow and rain, which reduced the snowfall accumulation.

through the Washington-Baltimore area. Within this band, 15 to 20 inches of snow fell, with 22 inches of snow reported in the northern suburbs of Baltimore.

Finally, during the late evening, the precipitation changed to heavy snow. The dense flakes fell rapidly throughout the rest of the night and into the morning of March 20. By noon, the worst of the storm was over, but periods of light-to-

The Snowstorm of March 19-20, 1958

The Snowstorm of March 19, 1958 began as a weak area of low-pressure just east of Norfolk, Virginia. Precipitation started in Washington as light rain. Unexpectedly, the storm strengthened and slowed its forward movement. The rainfall increased in intensity

A pole brought down by the weight of the wet snow, March 21, 1958. The precipitation began as rain, but changed to heavy, wet snow. The snow accumulated 4.8 inches in Washington, but ranged up to 33 inches in Mount Airy, Maryland. The liquid content of the rain and snow at National Airport was 3.75 inches. *Copyright Washington Post; Reprinted by permission of the D.C. Public Library*

moderate snow continued on into the morning of March 21.

The outstanding meteorological feature of this storm was the extreme amount of water content that fell. At National Airport, the water content of the rain and snow was 3.75 inches. Some local stations reported over 5 inches of liquid content.

With temperatures only marginally cold enough to support snow, the snowfall totals varied greatly with respect to location and elevation. At National Airport, where most of the snow fell with above-freezing temperatures, only 4.8 inches of wet snow accumulated. In Arlington, a foot of snow was measured. In the Maryland suburbs, 9 inches fell in Greenbelt; 11 inches fell in Silver Spring; 15 inches fell at Fort Meade; and 16 inches fell at Bethesda. Much of upper Montgomery County and Howard County received over 20 inches. At Mt. Airy, Maryland, 33 inches of snow fell. In the Baltimore area, snowfall depths also varied, from 8 inches at Friendship Airport (BWI) to over 2 feet in the elevated northwestern suburbs.

The sticky, wet snowfall clung to wires and poles, accumulating in snow rings of 3-4 inches in diameter. The weight of the snow proved too much, causing poles and wires to tumble throughout the area. The destruction of trees was also tremendous. Hundreds of trees were damaged or destroyed in Rock Creek Park alone.

At the peak of the storm, 300,000 area residents were without power and 10,000 were without phones. Approximately 800 power and phone lines were cut. In some areas, the power was not entirely restored for nearly a week. Civil defense officials were forced to gather generators for local hospitals and refuge centers were set up for those without power.

The Kennedy Inaugural Snowstorm of January 19, 1961

The Kennedy Inaugural Snowstorm began as a low-pressure center in northern Tennessee on the morning of January 19, 1961. It was starved for moisture and only produced light snow to the north of its track. By the afternoon of January 19, the storm center moved to Virginia and then "exploded" near the coast. Snow quickly developed in Washington.

When the snow began, the temperature was above freezing, in the mid-30's. By late afternoon, the mercury had dropped to the mid-twenties and the snowfall had become heavy. The flakes initially melted on the warm roadways, and then quickly froze into a glaze, which, in turn, was quickly covered by the wind-whipped snow. By evening, the rate of snowfall intensified and the winds increased from the northeast to 25 mph. National Airport reported visibility of zero – a total whiteout. Soon, commuters were abandoning cars and seeking shelter wherever they could find it. In what police called "a fantastic snarl," 288 vehicles were stranded on the George Washington Parkway between Alexandria and the Memorial Bridge and 200 others were stuck on Rock Creek Parkway. Over one hundred cars were abandoned on Pennsylvania Avenue.

Due to the storm, President-elect Kennedy was forced to miss the Eleanor Roosevelt reception and cancel his dinner plans. He and Jackie were later seen dashing in and out of Inaugural events amidst the swirling snowflakes.

President-elect John F. Kennedy and wife, Jacqueline, brave a snowstorm on their way to pre-inaugural events, January 19, 1961. The snowstorm ended before the Inauguration, but 7.7 inches of snow fell in Washington and created massive traffic problems.

Copyright Washington Post; Reprinted by permission of the D.C. Public Library

The snow tapered off by midnight, leaving 7.7 inches at National Airport. The liquid equivalent of the snow was 1.14 inches which is more typical of an 11-inch storm. Generally, about eight inches fell throughout the metropolitan area. Snowfall amounts were less to the south, with Richmond receiving one inch. Areas in northern Maryland received as much as 16 inches of snow.

The storm struck particularly hard throughout the Northeast. New England received

Top: The Kennedy Inaugural Parade on snow-covered roads near the Hotel Continental, January 20, 1961. Almost eight inches of snow fell during a vigorous, but short-lived storm that struck on January 19, 1961. *Copyright Washington Post; Reprinted by permission of the D.C. Public Library*

Left: Driving in Washington, February 3, 1961. Frigid, snowy weather gripped the D.C. area during the winter of 1961, with over 30 inches of snow measured at National Airport during the period from January 19 to February 12. *Copyright Washington Post; Reprinted by permission of the D.C. Public Library*

up to two feet of snow with 40 to 60 mph winds. In the Boston area, the air temperatures fell below 0°F at the height of the blizzard.

Snow removal crews worked feverishly throughout the night to clear Pennsylvania Avenue for the Inaugural Parade. By morning, Pennsylvania Avenue was in remarkably good shape. At noon, President Kennedy took the oath of office and gave his historic address in 22°F temperatures with a biting northwest wind. Despite the chill, over a million people lined Pennsylvania Avenue for the parade that followed.

The Kennedy Inaugural Snowstorm ushered in one of the most prolonged periods of subfreezing weather ever recorded in Washington. The temperature did not rise above freezing for eleven days – until January 31, when the high temperature reached 33°F. During that time, another 6-inch snowfall took place on January 26, which further enhanced the wintry conditions. The next month, February 1961, continued the cold and snowy trend, with more Arctic outbreaks and snowstorms.

The Ash Wednesday Storm of March 5-7, 1962

The famous Ash Wednesday Storm was noteworthy for producing devastating tidal flooding along the Atlantic Coast as well as record snows in the interior of Virginia. In Washington, the storm produced high winds and two separate periods of heavy, wet snow.

Along the Atlantic coast, tides ran 4 to 6 feet above normal with 20 to 40 foot waves crashing ashore. The extremely high tides and massive waves caused tremendous damage – worse than many of the hurricanes that had hit the region. Piers, boardwalks and beachfront property, including major hotels and apartment complexes, were severely damaged. Ocean City and Rehoboth Beach were among the communities that were devastated. Thousands of people from Florida to New York were left homeless. Total storm damage ran over $200 million and the death toll was 35, with three fatalities reported in Maryland and five in Delaware. President Kennedy declared a major disaster for both Maryland and Delaware.

To the south and west of Washington, extremely heavy snow fell, setting many snowfall records. Snow totals included: 15 inches at Richmond; 26 inches at Charlottesville; 32 inches at Winchester; and 35 inches at Front Royal.

Washington was spared the worst of the winds and snowfall. Snow began falling during the evening of March 5 and ended during the early morning of March 6. Then a long lull ensued. Moisture returned around the backside of the storm, causing snow to resume during the

Slushy snow at 12th and M Streets, NW after the Ash Wednesday Storm, March 7, 1962. Washington was spared the worst of the storm as the heaviest snow fell to the south and west of D.C. National Airport received 4 inches of snow with a liquid content of 1.33 inches. Front Royal, Virginia was buried under 35 inches of snow and Charlottesville, Virginia received 26 inches. At Ocean City, Maryland, winds gusted to 65 mph and waves of 30 feet pounded the shoreline. *Copyright Washington Post; Reprinted by permission of the D.C. Public Library*

FORECAST FOR DISTRICT OF COLUMBIA AND VICINITY

TUESDAY, MARCH 6, 1962

TODAY...WARNING HAZARDOUS DRIVING. SNOW MAY CONTINUE HEAVILY UNTIL AFTERNOON WITH MORE THAN 4 INCHES IN THE CITY AND UP TO 2 FEET IN THE WESTERN SUBURBS, TEMPERATURE IN THE 30'S.
TONIGHT...CLOUDY WITH LOWEST TEMPERATURE 25°.
WEDNESDAY...PARTLY SUNNY AND COLD.

The forecast issued for March 6, 1962. It was an accurate forecast for a difficult forecasting situation. National Airport received 4 inches of snow while Rockville, Maryland received 19 inches and Culpeper, Virginia received 23 inches. Many locations in western Virginia claimed new snowfall records, with Big Meadows, Virginia topping the list with 42 inches of snow. *NOAA Library*

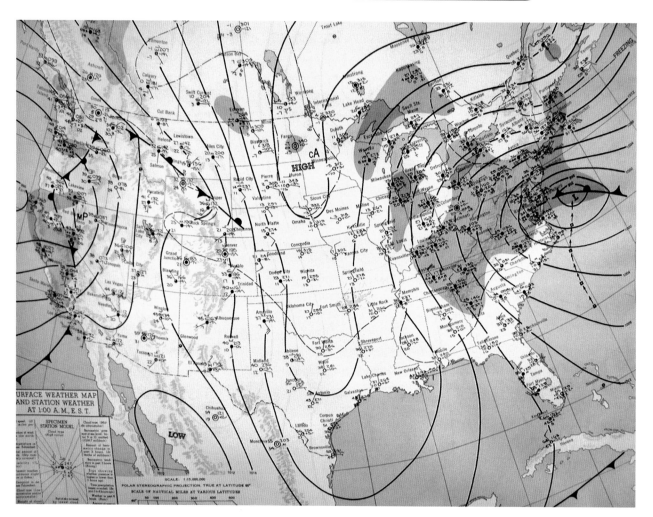

The surface weather map for March 6, 1962. The powerful Ash Wednesday Storm developed well off the Southeast coast and tracked northward off the Virginia coast, where it stalled. The central pressure dropped to 979mb and produced an easterly fetch of winds that exceeded 1000 miles, generating waves of 40 feet along the Middle Atlantic coast. *NOAA Library*

Walking home from school near 2nd and I Street SE, January 13, 1964. On the previous day, 8.5 inches of snow fell in Washington, with a high temperature of 22°F and a low temperature of 18°F.
Copyright Washington Post; Reprinted by permission of the D.C. Public Library

Screaming fans greet the Beatles at Union Station after venturing through a snowstorm, February 11, 1964. The snowstorm dropped 8.4 inches of snow on Washington and strong winds created 2-foot snowdrifts. The storm coincided with the arrival in Washington of the Beatles, who were on their first U.S. tour and were fresh from the Ed Sullivan Show. The snowfall did not deter 8,000 fans from later attending the band's performance at the Washington Coliseum. *Copyright Washington Post; Reprinted by permission of the D.C. Public Library*

Hamilton Street in Hyattsville, Maryland during the blizzard of January 30, 1966. The Blizzard of 1966 produced 13.8 inches of snow in Washington, with higher amounts to the south and east. The maximum sustained wind at National Airport was 37 mph, with the maximum gust clocked at 54 mph. Very cold temperatures accompanied the storm, with a high temperature of 16°F and a low temperature of 9°F on January 29. *Copyright Washington Post; Reprinted by permission of the D.C. Public Library*

afternoon which continued until midnight. In Washington, the storm was accompanied by strong northeast winds that were sustained at 30 to 40 mph, with gusts that approached 50 mph.

Approximately 200,000 homes in the area lost power due to the combination of high winds and heavy, wet snow. Southern Maryland was particularly hard hit by the outages. Fortunately, power was restored to all but 25,000 homes within 24 hours.

Much like the Snowstorm of March 1958, snowfall totals varied tremendously throughout the area. National Airport was on the low end of the range with only 4 inches. However, close-in suburbs, such as Silver Spring, Maryland and Falls Church, Virginia, received 11 inches of snow. Outlying areas such as Rockville, Maryland received 19 inches of snow and Leesburg, Virginia received 20 inches of snow.

The Blizzard of January 29-30, 1966

A fierce blizzard struck Washington on January 29-30, 1966. The storm's winds, clocked up to 54 mph, blew out plate glass windows and whipped snowdrifts up to ten feet high. It was the last of a series of coastal storms to hit the area in quick succession, leaving a deep blanket of snow over much of the Middle Atlantic area. The Blizzard of 1966 produced a swath of 12 to 16 inches of snow across Washington, falling on top of 3 to 6 inches of snow that was already on the ground. Frigid temperatures and high winds produced widespread blizzard conditions across the region.

The first storm occurred on January 22 and 23 and produced mainly rain in the immediate metro area, but 12 to 18 inches of snow fell in much of western Maryland and Virginia.

The second storm on January 26 produced heavy snows south and east of Washington. Richmond received 15 inches of snow and much of southern and eastern Maryland received 10 to 17 inches. National Airport reported 7.5 inches and Dulles reported 6 inches.

The third storm on January 29-30 was a true blizzard. A cold wave had surged in from the west shortly before a developing storm along the central Gulf Coast was moving northeast.

Car stuck in a drift on Wilson Boulevard in Arlington, Virginia, January 31, 1966. A series of snowstorms in January 1966 produced deep snow across the Middle Atlantic. The snow cover ranged from 16 inches at National Airport to 22 inches at Warrenton, Virginia. *Copyright Washington Post; Reprinted by permission of the D.C. Public Library*

The WWDC helicopter rescues an expectant mother in Prince George's County after the Blizzard of 1966. Helicopters from the military, police, and radio stations teamed together for emergency rescue missions since snowdrifts blocked many area roads. One snowdrift on Route 1 near Washington measured 12 feet high. *Copyright Washington Post; Reprinted by permission of the D.C. Public Library*

Cars buried in snow in Washington, D.C. after the Blizzard of 1966. National Airport received 13.8 inches of snow and Dulles Airport received 9.8 inches of snow. Snow totals were heavier south and east of Washington, with La Plata, Maryland receiving 14 inches of snow and Quantico, Virginia receiving 16.4 inches of snow.

Copyright Washington Post; Reprinted by permission of the D.C. Public Library

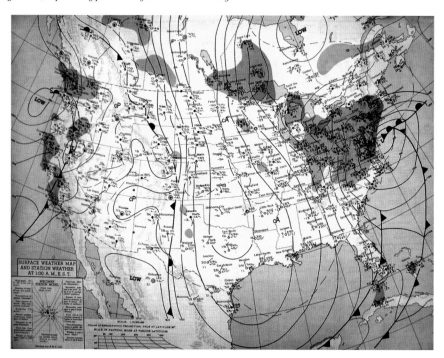

The surface weather map for January 30, 1966. A blast of artic air set the stage for the East Coast blizzard. The shaded area on the map represents precipitation. *NOAA Library*

During the early afternoon of January 29, the storm center was over eastern Georgia and light snow had started to fall in the Washington area. By evening, the storm moved to the South Carolina coast and snowfall in D.C. became heavy – falling at times at 2 inches per hour.

During the night of January 29-30, the storm curved to the north and intensified dramatically, deepening from 29.40 inches to 28.80 inches in just twelve hours. While the storm deepened, it also slowed down, stalling in the Delmarva area for about six hours. By morning, bona fide blizzard conditions were occurring across the metro area, with temperatures in the low teens and wind gusts over 50 mph.

Walking across the Memorial Bridge, February 7, 1967. The storm produced near-blizzard conditions and dropped between 10 inches and 12 inches of snow in the Washington area. *Copyright Washington Post; Reprinted by permission of the D.C. Public Library*

When the storm was over, National Airport reported 13.8 inches of snow. Overall, the Washington area received between 10 to 16 inches of snow. However, the effects of the storm were worse to the south and east of D.C. where snowfall depths ranged from 16 to 19 inches along the central Eastern Shore of Maryland. After the blizzard, up to 30 inches of snow was measured on the ground in the Delmarva area.

The Near Blizzard of February 1967

On February 7, 1967 a strong low-pressure system created near blizzard conditions in the Middle Atlantic region. The brief, but intense storm produced 10 to 12 inches of snow across the Washington area.

On February 6, a sharp cold front pushed through the area, ending a long mild spell. Behind the front, a large Arctic high-pressure system built

tures in the upper teens and winds of 20 to 30 mph, the snow blew around in a near blizzard-like fashion.

All schools were closed, but Federal employees were instructed to go to work that morning. Many stayed home, but those who made it to work had inched along a "nightmare course" of abandoned vehicles and wind-whipped snowdrifts. The Federal Government allowed the hardy commuters to leave two hours early that afternoon, but by then, the storm had ended.

The storm dumped 10.3 inches of snow at National Airport and 11.7 inches at Dulles. Generally, 10-12 inches fell throughout the area. The storm also hit the major Northeast cities with similar snowfall amounts.

The New Year's Day Snowstorm of 1971

A major snowstorm swept into Washington during the final hours of 1970 and continued into the first hours of 1971, disrupting most New Year's celebrations. Very heavy snowfall

in across New York and Pennsylvania. Meanwhile, a low-pressure system formed on the front near the Gulf of Mexico and moved northeast, eventually passing off the Virginia capes by the morning of February 7.

In Washington, light snow broke out during the evening of February 6. The snow remained light-to-moderate throughout the night, and then became heavy around daybreak. Snow accumulated at a 1 to 2 inch rate between 6:00 a.m. and 11:00 a.m., and then quickly ended by noon on February 7. With tempera-

A basket serves as a sled near Capitol Hill, March 3, 1978. The snowfall at National Airport measured 4.1 inches, with a high temperature of 35°F and a low temperature of 22°F. *Copyright Washington Post; Reprinted by permission of the D.C. Public Library*

totals were reported across the Washington area, with 8 to 16 inches of snow common in the immediate metro area.

The storm system tracked northeastward through North Carolina on the evening of December 31 and then slowly moved north, reaching Salisbury, Maryland early on New Year's Day. The snowfall in the metro area began during the afternoon of December 31, 1970. Snow continued to fall throughout the evening and ended by morning on January 1. At National Airport, 4.9 inches fell during the waning hours of 1970, and then an additional 4.4 inches accumulated after midnight, giving a storm total of 8.3 inches. The snow was driven by strong northeast winds, which reached 30 to 40 mph at times.

The storm struck particularly hard in the western suburbs, where Rockville, Maryland and Reston, Virginia reported 14 inches, and Dulles Airport tallied 15.4 inches. For Dulles, the storm set the airport's snowfall record, going back to 1963. The record stood until the Presidents' Day Storm of 1979.

In Washington and to the east, snow accumulations were reduced by sleet that fell for several hours. Only 6 inches of snow fell in Upper Marlboro. Distant western regions reported the heaviest snowfall, with 21 inches of snow in Cumberland, Maryland and 18 inches of snow in Winchester, Virginia.

The Presidents' Day Snowstorm of 1979

On February 18, 1979 a small, but intense low-pressure system "exploded" near Cape Hatteras, North Carolina and moved slowly up the coast. Snow began to fall in the District during the afternoon on February 18. On the morning of February 19 (Washington's Birthday), Washingtonians awoke to the biggest snowfall since the Knickerbocker Snowstorm of 1922. National Airport received 18.7 inches of snow, while up to 26 inches of snow buried the eastern suburbs. With 6 inches of snow on the ground before the storm, the snow cover in the Washington area ranged from 24 to 30 inches.

In the days preceding the storm, a bitter cold Arctic air mass had brought the coldest weather of the season to the Washington area, with highs in the teens and lows in the single digits. By February 18, a massive high-pressure system settled in over New York State. It was so expansive that its chilling influence was felt to the Gulf of Mexico.

Forecasters believed the storm would move south of Washington and out to sea, only grazing the area with a light snow of 1 to 3 inches. However, the storm intensified and moved north-northeast up the coast. As snow piled up across the Washington area, snowfall forecasts were updated frequently to

A woman throws a snowball at a friend who digs out a car in the 4400 block of MacArthur Boulevard NW, February 21, 1979. A snowfall of 18.7 inches on February 18-19, 1979 gave Washington its second largest snow on record. *Copyright Washington Post; Reprinted by permission of the D.C. Public Library*

A classic Georgetown snow scene on Wisconsin Avenue after the Presidents' Day Snowstorm, February 19, 1979.
A series of snowstorms dropped 30.6 inches of snow in Washington during a two-week period from February 6 to February 20, 1979. *Copyright Washington Post; Reprinted by permission of the D.C. Public Library*

Sailboats locked in ice and snow in Annapolis, Maryland after the Presidents' Day Snowstorm, February 19, 1979.
Two weeks of very cold temperatures had frozen portions of the harbor before the snowstorm dumped over twenty inches of snow on Annapolis. *Copyright Washington Post; Reprinted by permission of the D.C. Public Library*

Navigating through a sea of snow in Washington, February 20, 1979. Very heavy snowfall rates occurred during the Presidents' Day Snowstorm, with 4 inches of snow falling in one hour at Dulles Airport and 10 inches of snow falling in three hours in Baltimore, Maryland.

Copyright Washington Post; Reprinted by permission of the D.C. Public Library

Floral Street in Washington after the Presidents' Day Snowstorm, February 19, 1979. A very intense snowstorm hit Washington, starting on the afternoon of February 19, 1979 and lasting through the morning of February 20, 1979. The snowfall at National Airport measured 18.7 inches. The heaviest accumulations were to the east of the city, with 20 inches of snow falling in Baltimore, Maryland and 22 inches falling in Upper Marlboro, Maryland . *Copyright Washington Post; Reprinted by permission of the D.C. Public Library*

Snowmobiles in Georgetown after the Presidents' Day Snowstorm, February 19, 1979. Two weeks of extremely cold weather proceeded the snowstorm, culminating on February 18 with a high temperature of 15°F and a low temperature of 6°F. Warmer weather followed the storm and the snow quickly melted from roadways. *Copyright Washington Post; Reprinted by permission of the D.C. Public Library*

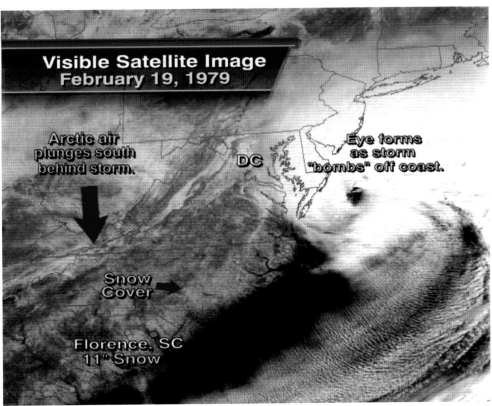

The visible satellite image shows that the Presidents' Day Storm has developed a well-defined eye, February 19, 1979. The storm quickly intensified off of the Middle Atlantic coast. *NOAA Library*

The surface weather map for February 19, 1979. The storm tracked out to sea, sparing Northeast cities from the heavy snowfall accumulations that occurred in the Middle Atlantic region. *NOAA Library.*

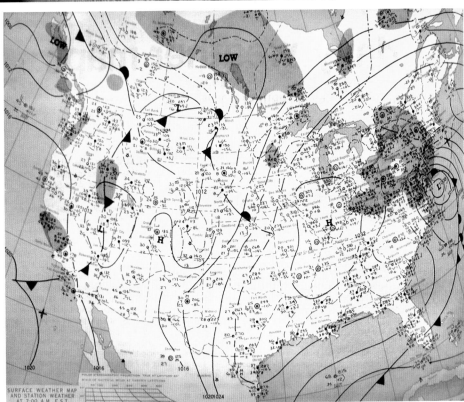

catch up with the rapidly increasing accumulations. By the morning of February 19, the forecast totals matched the actual snowfall amounts, but by that time the storm was winding down. National Airport received 18.7 inches of snow and Dulles Airport received 16.3 inches of snow.

Washington was paralyzed by the snowstorm. Coincidentally, U.S. farmers were in D.C. protesting for higher wages. They used their tractors to help city officials with snow removal. They also pulled cars out of snowbanks, cleared entrances to hospitals and fire departments and plowed parking lots. Farmers expanded their efforts to the suburbs and even delivered medication to snowbound individuals.

Soon after the storm, temperatures moderated and the snow began to slowly melt. However, many school systems remained closed the entire week.

The Record Early Snowstorm of October 10, 1979

Washington's earliest measurable snowstorm on record occurred on October 10, 1979. National Airport reported 0.3 inches of snow; however, much heavier snow fell to the west of Washington, causing significant tree damage in the mountains of Virginia. The tree damage was especially severe because the trees had not yet lost their leaves, allowing large amounts of snow to accumulate on their branches.

The setup for the storm began on October 9 when a low-pressure area moved east through New York State and Massachusetts. Washington was in the warm sector of the storm and temperatures topped out in the low 70's before the trailing cold front swept through during the late afternoon.

During the nighttime hours, unseasonably cold air surged down the East Coast. As the cold air was invading the D.C. area, a second storm center took shape over the Carolinas. A chilly rain broke out that evening and continued all night. By midnight, the temperature had fallen to 50°F.

The relentless drop of the mercury continued during the predawn hours and many people in the northern and western suburbs awoke to see snow falling. During the early morning, a burst of 1-3 inches of snow fell in central and northern Montgomery County and a coating of snow accumulated in Fairfax and lower Montgomery County.

The precipitation tapered off for a while in

all sections between 7:00 and 9:00 a.m., but by 10:00 a.m., a new band of heavy snow broke out, this time centering its fury on the southern half of the metropolitan area. Huge snowflakes were accompanied by lightning and thunder. By noon, the worst was over and the snow had tapered off.

During the second burst, 1 to 3 inches of snow fell in the central and southern parts of the region. A snowfall maximum of 3 inches was centered in Fairfax County and in the Upper Marlboro area of Prince George's County. Even National Airport, on the relatively warm Potomac, recorded 0.3 inches of snow, despite temperatures that never dropped below 36°F.

Snowfall amounts from this late morning burst rapidly decreased to the north – only flurries fell in the immediate northern suburbs, such as College Park and Silver Spring.

Approximately 114,000 customers in the area lost power during the storm. Trees were still fully covered with leaves at the time of the storm and were especially vulnerable to the heavy, sticky snowfall. Roads were mainly just slushy, but several schools were closed due to lack of heat.

Accumulating snow in October is a very rare occurrence in Washington. Aside from the October 10, 1979 storm, there have been only two measurable October snows on record. Those took place on October 19, 1940 and October 30, 1925. Both of those snowfall events were also in the 1 to 3 inch range.

The Snowstorm of January 13, 1982: The Air Florida Plane Crash

The snowstorm of January 13, 1982 will always be remembered for the terrible crash of Air Florida Flight 90. The storm was not very big – only four to eight inches of snow accumulated in the Washington area. However, the snowfall was heavy enough to accumulate on the wings of Flight 90 and contribute to its crash into the 14th Street Bridge.

The snow started lightly during the morning of January 13, but then intensified around noon. By early afternoon, roads in the area had become very hazardous as the snowfall

rate continued to increase. During a one-hour period, while the Air Florida flight was on the runway, the snow became very heavy. The visibility at National Airport briefly dropped to a sixteenth of a mile. Approximately 2 to 3 inches of snow fell during that hour. The snow ended abruptly, but by that time, the Air Florida jet had already crashed into the 14th Street Bridge and plunged into the Potomac River.

Prior to the storm, an extremely cold Arctic outbreak had spread across the eastern half of the U.S. Temperatures dropped to -25°F in Chicago and to near 0°F in Atlanta. There was also a major freeze in the central Florida citrus groves. At National Airport, the mercury dropped to 2°F. A second, even colder outbreak, invaded the area several days after the snowstorm and National Airport dropped to -5°F on January 17, with a high temperature of only 10°F.

The Snowstorm of February 10-11, 1983

On February 10-11, 1983 a huge snowstorm swept up the Eastern Seaboard, burying an area from Virginia to New York in a swath of very heavy snow. This was a textbook setup for heavy snow,

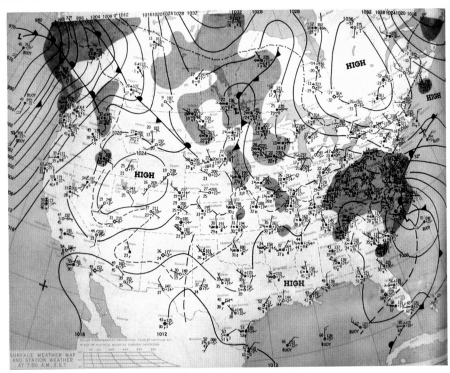

The surface weather map for February 11, 1983. The storm tracked northeast from Tallahasee, Florida to east of Wilmington, North Carolina, while Arctic high pressure remained anchored to the north. *NOAA Library*

AIR FLORIDA PLANE CRASH

Snow began to fall early on the morning of January 13, 1982. By midday, the snow was falling heavily and National Airport closed its runways for snow removal. About six inches of snow had accumulated in a fierce, but short-lived snowstorm. By mid-afternoon, the snow rapidly tapered off and the airport reopened. Air Florida Flight 90 was the first flight to be cleared for takeoff. The decision to fly was left to the pilot and he chose to takeoff even though a considerable amount of time had elapsed since the plane's last de-icing. At 3:45 p.m., the Boeing 737 jet taxied to National Airport's longest runway – a runway that was only 7000 feet with no overrun space at its end.

As Air Florida Flight 90 began to speed down the runway for takeoff, it struggled to gain traction and speed. Once the jet became airborne it never gained the proper speed or altitude for a safe takeoff. To complicate matters, the plane's takeoff was to the north, which requires a hard left turn over the 14th Street bridge to follow the course of the Potomac River.

The jet flew about one mile before it stalled. At 4:01 p.m., it slammed into the top of the 14th Street Bridge, crushing and scattering several vehicles. The 737 jet plunged over the bridge and into the ice-covered Potomac River below. The jet broke into multiple pieces as it shattered through the ice and then quickly disappeared into the river. In the icy water, about a dozen surviving passengers bobbed helplessly in the frigid water.

The tail from Air Florida Flight 90 sits aboard a recovery raft, January 19, 1982. The jet crashed into the Potomac River on January 13, 1982 during a snowstorm. The ensuing recovery effort was hampered by extremely cold weather, with temperatures plunging to –5°F on January 17, 1982.
AP/WIDE WORLD PHOTOS

A National Park Service helicopter arrived minutes later. The crew of the helicopter lowered a life preserver tied to the end of a rope and began to haul the survivors to shore. One passenger that was being dragged to shore lost her grasp of the lifeline and struggled in the water. Lenny Skutnik, an onlooker on the bank of the river, jumped into the icy water and swam out and rescued her. Another passenger, a man about 50 years old, passed the helicopter's life line several times to others, saving their lives. When the helicopter went back to get him, he had gone under.

Only five of the 79 on board Flight 90 survived the ordeal. Four commuters on the 14th Street Bridge were also killed. The total death toll for the crash was 78. Ice on the wings of the Air Florida jet was blamed as the cause for the crash.

Snow buries a neighborhood near Frederick, Maryland, February 12, 1983. Frederick, Maryland received 30 inches of snow. The maximum snowband associated with the storm stretched over Washington's western suburbs and extended west to the mountains. National Airport received 16.6 inches of snow; Manassas, Virginia totaled 22 inches of snow; and Damascus, Maryland measured 24 inches of snow. *Katie Kahan*

Sunrise over the water at Chesapeake Beach, Maryland, February 1983. Sleet and snow pellets reduced snowfall accumulations to the south and east of Washington. Between eight to twelve inches of snow fell in the southern portion of Calvert County, Maryland. Farther east, Salisbury, Maryland reported only 5.3 inches of snow and sleet. *Joseph Reintzel*

with a nearly stationary high-pressure area sitting north of New York in a position to provide cold air to the Washington area. At the same time, a moist low-pressure system had developed and was moving northeast from Alabama.

During the late evening of February 10, with the low over Georgia, light snow broke out in the Washington area. The snow slowly picked up during the overnight hours. By the morning of February 11, the surface low was positioned just east of Wilmington North Carolina. At that time, heavy snow was falling across the entire region. A tightening pressure gradient between the storm and the high-pressure area to the north caused northeast winds to increase, with gusts over 40 mph.

Most locations reported the heaviest snowfall rates, commonly 3 inches per hour, during the late morning and again during the mid-to-late afternoon of February 11. The snow tapered off by evening. Many observers, particularly in the Maryland suburbs, reported several episodes of lightning and thunder. The snow thunderstorms along the Eastern Shore produced extremely strong winds and whiteout conditions.

Most of the southern and eastern suburbs recorded 15 to 20 inches while 20 to 30 inches fell in the northern and western suburbs. National Airport received 16.6 inches while 22.8 inches fell at both BWI and Dulles Airport.

In northwest Montgomery and Frederick Counties, the storm was the greatest ever recorded, easily exceeding

their totals received during the famous Knickerbocker Storm of January 1922. Germantown and Frederick both received 30 inches of snow. In western Loudoun County, Virginia, up to 38 inches of snow fell. Likewise, Braddock Heights, just west of Frederick, Maryland, received 34.9 inches.

From Washington east, snow pellets mixed in with the snow. That reduced the accumulations, but made for a considerably denser snow pack.

The "Double Whammy" Snows of January 22 and 25-26, 1987

Going into the fourth week of January 1987, the winter had been mild and almost entirely free of snow in the Washington area. That pleasant state of affairs ended abruptly in late January when two back-to-back snowstorms dropped 20 to 28 inches of snow across the D.C. area.

On the morning of January 22, a low-pressure system was located along the South Carolina coast. It quickly moved north-northeast while developing into a major winter storm. During the afternoon of January 22, the storm's

Deep snow near Leesburg, Virginia, January 27, 1987. Between 20 to 30 inches of snow fell in back-to-back snowstorms on January 22 and January 25. Snow cover after the storms ranged from 16 inches at National Airport to 24 inches in Charlottesville, Virginia. *WJLA*

Leesburg Airport is buried by 20 inches of snow, January 27, 1987. The winter of 1986/1987 had been relatively snow-free before the back-to-back snowstorms of January 22 and January 25. A month later, on February 22-23, another major snowstorm hit the area with 14 to 16 inches of snow accumulating in the northern and western suburbs. *WJLA*

central pressure dropped below 29 inches as it passed just east of the Virginia coast.

Snow became heavy in Washington during the morning and continued at a relentless 1 to 3 inches per hour until mid-afternoon. Between 10 to 15 inches of snow fell throughout the entire metro area. National Airport received 10.8 inches, BWI received 12.3 inches and Dulles received 11.1 inches. The heaviest totals were concentrated in northwest D.C. and the immediate northern and western suburbs. The snow had an unusually high water content, 1.49 inches of liquid equivalent at National Airport, which made snow removal difficult.

Temperatures remained cold on January 23-24, with almost no melting of the snow cover. By Super Bowl Sunday, January 25, a new storm center took shape over Mississippi as an Arctic high-pressure system ridged from south-central Canada to Pennsylvania. That frigid air mass kept Washington's temperatures from climbing above a high of 17°F.

Light snow broke out on the morning of January 25 and continued throughout the day, making area roads hazardous. The snowfall became heavy during the overnight hours, and

then tapered off during the morning of January 26.

This time, the storm took a more easterly track, passing over Cape Hatteras and moving farther offshore than its predecessor did. That track put Washington near the northwest edge of the heavy snow band. Generally, 9 to 14 inches of new snow fell in the metro region, with the heaviest amounts to the south of D.C. National Airport received 9.2 inches, Dulles received 10.1 inches and Baltimore, Maryland received 9.6 inches. Patuxent and Quantico topped the list with a whopping 16 inches. The combined snow cover across the Washington area after both storms was between 16 to 24 inches – National Airport had 16 inches; Baltimore, Maryland had 17 inches; Dulles Airport had 18 inches; and Charlottesville, Virginia had 24 inches.

Washington, D.C. Mayor Marion Barry had left town on Wednesday, January 21 to attend the Super Bowl in Pasadena, California. He returned on the night of Monday, January 26 after basking for five days in 80-degree weather. By then, the Mayor was receiving criticism by snowbound residents who had been unable to drive on their unplowed streets. The perception was the suburbs were doing much better with snow removal than the city, a contention vehemently denied by D.C. officials. From that time forward, snow removal became a mayoral campaign topic in Washington, as it already was in many northern cities.

Veterans' Day services in the snow at Arlington, Virginia, November 11, 1987. A snow of 11.5 inches fell at National Airport during the late morning and afternoon of November 11. There was a tremendous snowfall gradient associated with the storm – Washington's western suburbs received 3 to 5 inches of snow while eastern suburbs of Washington had up to 17 inches of snow. *WJLA*

The Veterans' Day Snowstorm of November 11, 1987

The Veterans' Day snowstorm of November 11, 1987 was by far the heaviest snowstorm to hit the Washington area so early in the season. The 11.5 inches that fell at National Airport easily broke the old November record of 6.9 inches that fell on November 30, 1967. The next earliest date for a snowstorm of that magnitude occurred well into the month of December when 12 inches of snow fell on December 17, 1932.

National Airport was in a heavy snow band that reached its maximum of 14 to 16 inches in western Prince George's County. Snowfall amounts tapered off rapidly to the northwest, with Gaithersburg reporting only 3 to 4 inches of snow.

As the storm headed northeast, up the East Coast, the quirky snowfall pattern continued. Philadelphia and New York City reported virtually no snow, but both Boston and Providence received 10 inches of snow, also setting early season records.

On the weekend preceding the storm, November 7-8, the region experienced beautiful 70°F weather. On Monday, November 9, a cold front brought an abrupt end to the mild spell.

The snow began during the early morning of November 11 with a quick burst that produced 1 to 2 inches across Washington. That snowfall quickly ended, leading many to believe that the storm was over. For a few hours in the morning, no snow fell and people headed off to work, school and shopping.

During the late morning, ominous streaks of heavy snow began to break out south of D.C. The heavy snow surged northward as a second low-pressure system quickly intensified. By noon, a very heavy, nearly stationary, band of snow had set up through the eastern half of the area. Localized within this band, snow fell at a whiteout rate of 3 to 4 inches per hour for several hours, accompanied by frequent lightning and thunder. Outside that band, only light-to-occasionally-moderate snow fell. Temperatures hovered around 30°F and since the texture of the snow was not particularly wet, power outages were not a problem. When the snow stopped in the evening, the snow accumulation in the metro area ranged from 3 to 16 inches.

The end of the snowfall did not signal the end to traffic tie-ups. A very large backup occurred on Interstate 95 south in Prince William County and another fifteen-mile backup stacked up on the Beltway. An estimated 60,000 travelers were stuck in traffic for much of the evening. Some commuters spent the night in their cars while others abandoned their vehicles and found shelter elsewhere.

Following the storm, temperatures rapidly moderated. On November 12, a high temperature of 48°F allowed for a quick return to normal road conditions. By the weekend of November 14-15, the Indian Summer weather had returned with temperatures near 70°F.

The Storm of the Century, March 13, 1993

The "Super Storm" of March 13, 1993 will go down in history as one of the largest winter storms on record. Heavy snow and blizzard conditions extended from the Gulf States to New England and from the Ohio Valley to the East Coast. Hurricane force winds battered cities along the Atlantic Coast. Deadly tornadoes were spawned in Florida and tremendous waves and tides occurred from Key West to Maine. The storm was so large that its effects were felt from Cuba, where high winds and rain damaged the sugar crop, to Chicago, where 250 flights at O'Hare International Airport were grounded due to snow squalls. Approximately 270 deaths were attributed to the storm, three times that of Hurricanes Andrew and Hugo combined.

The storm originated as a cluster of thunderstorms over Texas on the morning of March 12. As the late winter storm moved over the warm Gulf of Mexico waters, a storm of hurricane proportions began to take shape. Many buoys in the central Gulf recorded wind gusts over 100 mph. A cluster of tornadoes hit Florida, while at the same time an 11-foot tidal surge hit the west coast of Florida during the night of March 12-13. A total of 44 people died in the Sunshine State.

The highest recorded wind gust associated with the storm occurred on Mount Washington, New Hampshire where the wind was clocked at 144 mph. Dry Tortugas in Florida (west of Key West) recorded a 109-mph wind gust; Myrtle Beach, South Carolina recorded a 90-mph wind gust; and Fire Island, New York recorded an

The Storm of the Century
Infrared Satellite Imagery: 3/12/93

Ski-sailing by the Washington Monument, March 14, 1993. Six to fourteen inches of snow fell across the Washington area on March 13. The snowfall at National Airport measured 6.6 inches and the snowfall at Dulles Airport totaled 14.1 inches. Sleet and rain mixed with the snow in Washington, which reduced snowfall accumulations. The peak wind gust at National Airport clocked in at 47 mph, from the northeast. *AP/WIDE WORLD PHOTOS*

The surface weather map for March 13, 1993. The map shows the "Superstorm" poised to move up the Eastern Seaboard.
NOAA Library

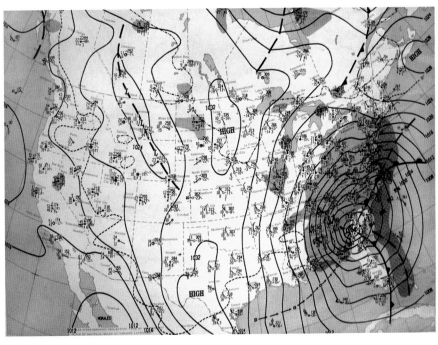

89-mph wind gust.

Incredible snow totals occurred with the storm, including 50 inches at Mount Mitchell, North Carolina; 43 inches at Syracuse, New York; 30 inches at Beckley, West Virginia; 25 inches at Pittsburgh, Pennsylvania; 20 inches at Chattanooga, Tennessee; 15 inches at Birmingham, Alabama; and 14 inches at Washington Dulles Airport.

Record low pressures accompanied the storm. All-time record low pressures include 28.38 at White Plains, New York; 28.54 inches at Washington; 28.64 inches at Columbia South Carolina; and 28.86 inches at Tallahassee, Florida.

During the evening of March 12, heavy snow broke out from Alabama through the western Carolinas. The snow spread rapidly northeast, reaching D.C. shortly after midnight. It became moderate-to-heavy during the morning. By then, most of the area had received 4 to 8 inches of snow.

During the late morning, sleet began to mix with the snow, especially from Washington south and east. By early afternoon, many areas experienced rain. By evening, the precipitation had changed back to snow. The snow was accompanied by high winds, with a peak wind gust at National Airport recorded at 47 mph.

The period of mixed precipitation in the Washington area kept accumulations down, with 7 to 12 inches common in D.C. and the close-in suburbs. At National Airport, 6.6 inches of snow fell; at Dulles, 14.1 inches of snow fell; and at BWI, 11.9 inches of snow fell. The heaviest accumulations were generally west of the city, but there were also pockets of 12-plus inches of snow

to the southeast and northeast of the District. Generally, the liquid totals for the storm were in the 2 to 3 inch range.

The National Weather Service is credited with issuing timely, accurate warnings that undoubtedly saved many lives. The forecast for a major storm had been issued several days before the storm had even developed.

The Great Sleetstorm of February 10-11, 1994

One of Washington's greatest sleetstorms in history occurred during the extremely cold winter of 1994. The winter of 1994 featured very little snow; however, a series of ice storms hit the Washington area during the months of January and February. The winter storm of February 10-11 was unique in that it produced almost entirely sleet for the immediate D.C. metro area. To the north, heavy snow fell and to the south and east, freezing rain occurred, producing a devastating ice storm.

In the Washington area, the sleet accumulation averaged between 3 to 4 inches. Temperatures hovered in the low-to-mid twenties

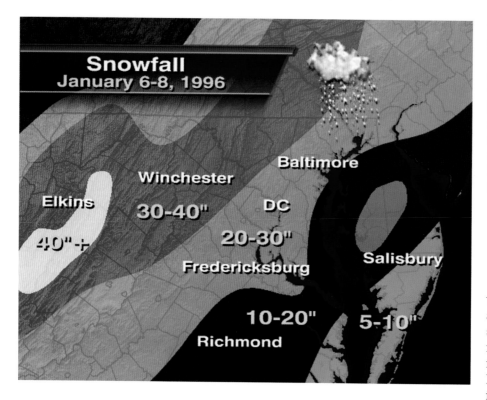

Snowfall
January 6-8, 1996

Elkins
40"+

Winchester
30-40"

Baltimore

DC
20-30"

Fredericksburg

Salisbury

10-20"
Richmond

5-10"

the 1927 storm was 4.5 inches.

To the south and east of Washington, where significant freezing rain fell, there was extensive tree damage and widespread power outages. Cleanup from the storm lasted for months in those regions.

The Blizzard of January 7-8, 1996

The Blizzard of 1996 was incredibly massive and truly historic in its scope. All-time snowfall records were widespread, including 24.9 inches in Roanoke, Virginia; 30.7 inches in Philadelphia, Pennsylvania; 27.8 in Newark, New Jersey; and 14.4 inches as far away as Cincinnati, Ohio. Every city along the northeast megalopolis, from

during the event and the sleet was very dry, accumulating on the ground in what resembled snow. However, after walking or driving in the sleet, it became very apparent that the accumulation was not light or powdery like snow, but rather coarse and heavy like sand.

The maximum sleet accumulation occurred over central and western Fairfax County, with over 4 inches of sleet measured at several locations. Only two other sleetstorms in Washington's history have compared to this storm – one occurred in 1920 and the other in 1927. The average sleet depth for

The color-enhanced infrared satellite image for January 7, 1996. The Blizzard of 1996 is in its formative stages in the Gulf of Mexico. *NOAA Library*

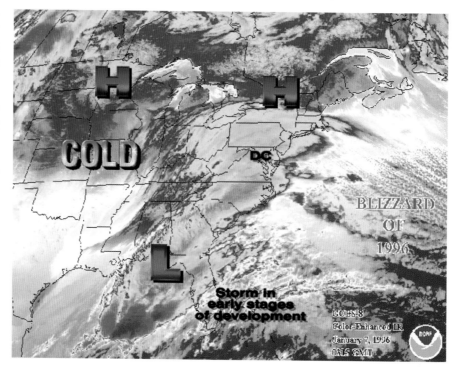

Top: A car is almost completely buried by snow near Fairfax, Virginia after the Blizzard of 1996, January 8, 1996. The storm dropped heavy snow from Washington to Boston. The snowfall at National Airport measured 17.1 inches while the snowfall at Dulles accumulated 24.6 inches. Other snowfall totals included: 21 inches in Fredericksburg, Virginia; 22.5 inches in Baltimore, Maryland; and 25.7 inches in Rockville, Maryland. *Kevin Ambrose*

Right: A parking space in Fairfax, Virginia is marked "Cleared & Occupied," January 13, 1996. Car owners protected parking spaces after digging their cars out from under two feet of snow. Cold temperatures and two moderate snowfalls that followed the Blizzard of 1996 helped maintain a deep snow cover across the Washington area for more than a week.
Kevin Ambrose

Snowdrifts near the reflecting pool, January 14, 1996. The maximum wind gust in Washington during the Blizzard of 1996 clocked in at 39 mph, from the north. *Kevin Ambrose*

A John Deere tractor used to dig a path to the Lincoln Memorial after the Blizzard of 1996, January 14, 1996. Much of the Washington area remained buried by snow a week after the blizzard. *Kevin Ambrose*

Washington to Boston, received between 17 to 30 inches of snow. While other storms have been bigger in any given city, it appears that there has never been a greater snowfall event for such a large, highly populated area of the Eastern U.S.

Snow began falling in Washington during the late evening of Saturday, January 6 and continued at an amazingly steady rate until mid-afternoon Sunday, January 7. By that time, 13 to 17 inches of snow had accumulated in most areas, with up to 20 inches in the distant western suburbs.

For several hours during the afternoon, sleet took over, primarily in the eastern suburbs. The precipitation stopped during late Sunday afternoon and early evening, giving the impression that the worst of the storm was over.

But this proved to be just a lull. During the overnight hours, as the storm crawled slowly north along the Delmarva Peninsula, bands of heavy snow redeveloped and swept into the area from the east. The snow bands were accompanied by lightning, thunder and whiteout conditions. Gusty north winds, which at times approached 40 mph, created a blizzard-like scene. By Monday morning, January 8, the snow squalls had tapered off, leaving the Washington Metro area buried under 15 to 25 inches of snow.

The snowfall ranged between 15 to 20 inches in the southeastern portion of the metro area and 20 to 25 inches in the northern and western suburbs. National Airport received 17.1 inches of snow, Dulles Airport received 24.6 inches of snow and BWI received 22.5 inches of snow. In nearby northern and western suburbs, Silver Spring, Maryland received 23 inches of snow and McLean, Virginia received 22 inches of snow. Totals were even heavier farther to the north and west, with Rockville, Maryland receiving 25.7 inches of snow

and Damascus, Maryland receiving 30 inches of snow.

On Monday, road crews began the epic struggle to dig out from the crippling snowfall. The entire region was shut down. High winds caused snow to blow back onto plowed roads, and many streets were not plowed at all. The Metro system fared no better as frozen rails crippled the outside portion of that system. One train, with one hundred passengers on board, got stuck for five hours near Takoma Park.

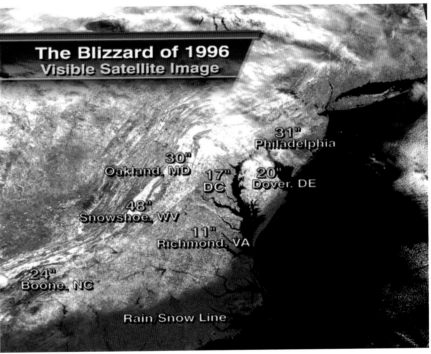

The visible satellite image shows the snow cover after the Blizzard of 1996. The location of the rain/snow line is clearly visible through southeast Virginia and central North Carolina. *NOAA Library*

The Blizzard of 1996 was just the first of three snowstorms to hit the Washington area during the snowy week of January 7-12, 1996. On January 9, a weak Alberta Clipper storm center passed directly over the region, dropping another quick shot of surprisingly heavy snow on the area. Transportation officials across the region were forced to refocus their efforts on the major highways.

A snowman on the Mall near the Smithsonian, January 14, 1996. The snowman stood until the rapid thaw of January 19, when temperatures reached 62°F and heavy rain fell across the area. *Kevin Ambrose*

The Alberta Clipper snowfall pattern was very different from the blizzard. The western suburbs were dusted with only an inch of snow while up to six inches of snow fell in the eastern areas such as Prince George's and Charles Counties. At the conclusion of the Clipper storm, two-foot snow depths were fairly common across all sectors of the metro region.

Wednesday, January 10, was a nice sunny day with the high temperature reaching a relatively balmy 34°F. Nevertheless, snow removal in residential neighborhoods proceeded quite slowly. As late as Wednesday afternoon, fifty percent of Prince George's County residential streets were not plowed and forty percent of Fairfax County's streets had not seen a plow.

Finally, on Thursday, January 11, the Federal Government reopened for the first time in nearly a month, following an extended furlough and the weather-related shutdown. However, the morning and evening rush hours proved disastrous, with major roads narrowed significantly by tall snowbanks. Also, many Metro trains remained out of commission, still mired in deep snowdrifts.

The third and final blow of the snowy sequence took place on Friday, January 12 as a quick moving coastal storm dropped 5 to 12 inches of snow across the region. The snow began during the predawn hours and was over by early afternoon. Snow fell heavily during midmorning, before switching over to sleet east of Washington. West of D.C., the storm produced mostly snow and was heaviest in central Maryland. From Philadelphia northeastward to Boston, the storm produced mostly rain, with only a brief period of snow before a changeover to rain occurred.

The Federal Government was once again closed on Friday, January 12, as were most schools and business establishments. Fortunately, temperatures warmed into the 40's during the three-day Martin Luther King Holiday weekend of January 13-15. Finally, on Tuesday, January 16, schools and businesses reopened, after an 11-day blizzard holiday.

However, many problems still remained. As late as Tuesday, January 16, dozens of Washington neighborhoods had still not seen a snowplow and were virtually impassable to traffic. The commute on Tuesday, January 16, was terrible. Snowbanks blocked traffic lanes and sidewalks, causing drivers to battle pedestrians for the right of way. Some commuters spent half of the day simply getting to and from work.

A rapid thaw brought 62°F temperatures and heavy rain into the area on Thursday, January 18. The huge snow piles vanished overnight and green grass appeared where deep snow had blanketed the ground just a day earlier.

The sudden meltdown sent the Potomac River surging far out of its banks. The river approached levels not seen since Tropical Storm Agnes in 1972. Several blocks of Old Town Alexandria were under water and the George Washington Parkway, Clara Barton Parkway, and many other roads were closed due to flooding.

The Ice Storm of January 14-15, 1999

A destructive ice storm struck the Washington area on January 14-15, 1999, causing one of the greatest power outages in recent history. An inch of freezing rain caused numerous trees, limbs and power lines to fall throughout the area.

Heavy ice accumulations on tree limbs near Chantilly, Virginia, January 15, 1999. A shallow layer of subfreezing Arctic air covered the Washington area while temperatures several thousand feet above the ground were above freezing. Rain fell through the warmer air aloft and froze at the surface, coating everything in a thick glaze of ice.
Kevin Ambrose

A utility pole and large tree in Fairfax County snap under the weight of heavy ice, January 15, 1999. Nearly a half-million people in the greater Washington area lost power during the ice storm. *WJLA*

major damage. To the east of Washington, in Prince George's and Anne Arundel Counties, the ice damage was not nearly as severe. The eastern counties had slightly warmer temperatures and less ice accumulation than neighboring counties to the north and west.

The Unusual Snowstorm of March 9, 1999

A massive Arctic high-pressure area, with a central pressure around 31 inches, was located north of New York State early on January 14. In parts of New York and New England, temperatures did not rise above zero the entire day. A shallow layer of frigid air from the Arctic high drained south along the east side of the Appalachian Mountains, causing surface temperatures to drop below freezing in Washington. Several thousand feet above the ground, temperatures were above freezing, setting the stage for an ice storm

Light freezing rain fell throughout the day on January 14. Overnight, the ice storm intensified and the freezing rain began to fall rather heavily. Surface temperatures stayed well below freezing in most areas, ranging from the upper twenties at Reagan National to the upper teens in the far northern and western suburbs. During the predawn hours, the weight of the ice snapped thousands of tree limbs, which took down numerous power lines. In Montgomery County alone, 1223 wires were counted down and 11 substations were knocked out of commission. On January 15, 435,000 local customers throughout the area were without power!

Montgomery County was hard hit by the ice storm. Likewise, Fairfax County also suffered

On March 9, 1999, a storm over the Ohio Valley produced a very heavy band of snow that streaked directly across the Washington area. As a general rule, Washington's biggest snowstorms occur when moist storm systems track northward from the Southeast U.S., or when coastal storms develop near the Virginia or North Carolina coast. However, the storm of March 9, 1999 was an exception to this rule.

In the early morning of March 9, 1999, a strong high-pressure system was helping to drive very cold, dry air southward into Virginia. Meanwhile, a rather weak storm was moving northeast through the lower Ohio Valley. Indications were that the dry air mass over the region would keep accumulations minimal. However, a very narrow band of heavy snow in central and northern Ohio during the night of March 8–9 moved rapidly towards Washington during the predawn hours.

On the morning of March 9, snow began falling in Washington and the western suburbs. This band expanded eastward and became almost stationary, lasting through the afternoon. The heavy snow band was oriented east-to-west, and was centered just a few miles to the south of D.C. The heaviest snow fell in central Fairfax County, where 10 to 12 inches accumulated. On average,

The visible satellite image shows the narrow stripe of snow cover across the Washington area, March 10, 1999. A narrow band of snow set up directly over the Washington area as a storm moved southeast from Ohio. Up to a foot of snow fell in Fairfax County. *NOAA Library*

The Nor'easter of January 25, 2000

On January 24-25, an unexpected nor'easter hit North Carolina and Virginia. Between 8 to 18 inches of snow fell across the Washington area. The heaviest totals were to the east and south of Washington. The storm was a record-breaker in North Carolina, where Raleigh tallied over 20 inches of snow.

8 to 10 inches of snow fell in the immediate Washington area. Reagan National Airport reported 8.4 inches of snow – the heaviest March snowfall since the storm of March 28-29, 1942. Snowfall amounts tapered off significantly to the north and to the south of Washington. The north side of Baltimore received only 2 inches of snow, while Charlottesville and Richmond received light accumulations.

As late as the afternoon of January 24, it appeared that the developing storm system in the southeast U.S. would track out to sea, south of Washington, sparing the area of significant snow accumulation. Dry air had surged into Virginia

This was not a typical, wet March snow. With high temperatures in the middle 20's during the afternoon, the snow had little trouble sticking to roadways. The snow was also quite dry. However, there was almost no wind associated with the storm, so drifting was not a factor.

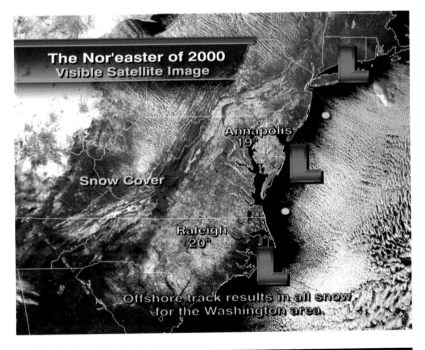

The visible satellite image a day after the Nor'easter of 2000 showing the heavy snow cover over North Carolina, Maryland and Virginia, January 26, 2000. New England was spared the worst of the storm while Raleigh, North Carolina received 20 inches of snow. *NOAA Library*

from the north and west, and the skies cleared in the afternoon, which produced a striking, red sunset. However, later in the evening of January 24, it became quickly apparent that the storm was going to take a track up the East Coast and not out to sea. The precipitation that had stayed focused in the Carolinas during most of the day was now moving northward into Virginia. Also, precipitation was advancing westward from the Atlantic Ocean into Eastern Virginia and Maryland. The storm was rapidly strengthening near the coast of North Carolina and all East Coast cities were quickly put on alert for winter storm conditions.

On January 24 at 9:07 p.m., the Weather Service issued a Winter Storm Warning for the Washington area. Minutes later, the television networks interrupted programming with emergency weather bulletins. The late night news programs focused on the impending storm; however, many people had missed the warnings.

The snow began during the early hours of January 25, and within an hour had become moderate-to-heavy. As the storm moved up the coast, snow bands set up across the area with a north-to-south orientation. The heaviest snow band was located in the eastern suburbs, with another band located well west of D.C. The storm continued with varying intensity throughout the day and did not end until the evening. The snow was accompanied by high winds that gusted past 40 mph, which caused considerable drifting.

Snow totals ranged from 8 to 12 inches from Washington to the west, with generally 12 to 18 inches of snow east of D.C. The deepest snow was measured near the county line of Prince George's and Anne Arundel Counties. Snow totals were 9.3 inches at Reagan National Airport, 10 inches at Dulles Airport, 14.9 inches at BWI, and 18 inches at Annapolis, Maryland.

Another storm developed the following weekend and again moved up the coast. This time, the precipitation was not nearly as heavy and included sleet and freezing rain – especially east of Washington. Snow totals with the second storm ranged from 1 to 3 inches to the east of D.C. and 4 to 9 inches in the far northern and western suburbs.

Large snowbanks in Indian Head, Maryland after the Nor'easter of January 25, 2000. The snowfall in Washington tallied 9.3 inches. Farther east, snow totals increased, with 14.9 inches in Baltimore, Maryland and 19 inches in Annapolis, Maryland. *John Olexa, Jr.*

Snow falling in the forest, Upper Marlboro, Maryland, January 25, 2000. The heaviest snow fell in the eastern suburbs, where up to 19 inches accumulated. *Marcia K. Hovenden*

Playing in the snow during the Nor'easter of January 25, 2000, Oakton, Virginia. Between eight to ten inches of snow fell across the western suburbs during the storm, while the eastern suburbs tallied up to 19 inches. *Kevin Ambrose*

Mattawoman Creek in Charles County, Maryland, after the Nor'easter of January 25, 2000. Cold temperatures froze many area waterways. *John Olexa, Jr.*

SLUSH FUNDS - A HISTORY OF D.C. SNOW MANAGEMENT
BY BERNARD MERGEN

With an average of approximately 17 inches of snow annually, Washingtonians might well ignore winter. Cities such as Buffalo, Rochester, and Syracuse, New York, average more than five times this amount, Boston triple, New York, and Philadelphia double. Yet snow is important to everyday life in the nation's capital and to its image.

Even 17 inches of snow falling over three months in a dozen storms can significantly affect people's lives. The failure of Congress, city administrators, and residents themselves to recognize the effects of snow offers an interesting case study of environmental naivete'. As the geographer John Rooney, Jr., has pointed out, it is cities that fail to confront the "urban snow hazard" that suffer more than cities with greater snowfall but that are well prepared.

Snow reduces cities to earlier historical times. A heavy snowfall confines or eliminates automobile traffic and strips bare the shelves of supermarkets. The transient nature of much of metropolitan Washington's population – tourists, political sojourn-

ers, recent immigrants from tropical zones – also contributes to the capital's unique responses to snow. Situated on the edge of the snow frontier separating the snow-covered North from the snowless South, the city is a mecca for nivaphobes and nivaphiles alike, the former lamenting their inability to control the weather, the latter delighting in their participation in an American experience. Daily reports on television, radio, and in newspapers remind us of our national as well as local weather.

Snow has always been seen as a kind of litter in the streets, refuse for the sanitation department to remove. Nature's bounty is also nature's toll, a paradox that neither funds for slush, nor slush funds can solve. Below is the chronology of how this has evolved in DC over the past one hundred years:

1895 – Congress passed an act that required "the owner or tenant of each house or other building, or lot or lots of ground in the cities of Washington and Georgetown" to remove, "within the first four hours of daylight after every fall of snow, all the snow from the sidewalks opposite their premises, and to strew them "with ashes, sand, sawdust, or some suitable substance that will insure or contribute to the safety of pedestrians." Those who failed to comply could be fined one dollar, or the commissioners could assess a special property tax to cover the cost of snow removal.

1897 – An act was passed that extended responsibility for sidewalk snow removal to real estate agents selling unoccupied property.

1899 – After the Blizzard of 1899, the Senate quickly responded to the urgent request from the District Commissioners appropriating $20,000 to pay for the 2,000 men and 500 carts needed to remove the snow. The House took another day to follow suit, many members complaining that their constituents should not have to help pay for cleaning Washington's streets. The responsibilities of the private streetcar lines and homeowners to remove snow from their properties were left unresolved.

1904 – The 1895 act was repealed by the act of February 10, 1904, which retained the responsibility of tenants and owners for snow removal from sidewalks adjacent to their property and increased the penalty to a five-dollar fine or five days in jail. The commissioners were required to remove snow in front of public buildings and squares, but no money for this purpose was appropriated. Within three months this act was challenged and held unconstitutional by the Court of Appeals.

1905 – The District's commissioners attempted, but failed, to assert their authority to compel owners to clear snow from pavement adjacent to their property. In Coughlin v. District of Columbia, decided March 22, 1905, the Court of Appeals declared that only Congress had the authority to make regulations for the removal of snow.

1910 – The D.C. commissioners appealed for snow removal legislation in 1910, noting that it would require 12,000 men working four hours to clear away a three-inch snowfall from the city's estimated 550 miles of paved sidewalks, but Congress failed to act. Seven years later the commissioners estimated that a six-inch snowfall would require 50,115 men and 19,162 teams. The Street Cleaning Department employed 350 men and 90 teams.

1922 (January) – After the snowstorm of January 27-28, the president of the Federation of Citizens' Associations appealed to all citizens to do their duty and clear snow from in front of their premises. Streets blocked with snow were blamed for delaying emergency vehicles. Morris Hacker, superintendent of the street-cleaning department told reporters that he had only 200 men for snow shoveling but planned to recruit 450 more. Senator Dillingham of Vermont blamed members of the House District Committee for pigeonholing his bill to compel District householders to remove snow from in front of their properties.

1922 (February) – A bill similar to that of 1895, but one that allowed property owners up to eight hours to remove snow, passed February 6, 1922, and technically remains in effect, although a personal injury suit brought in 1948 when a boy slipped on the uncleared sidewalk in front of a private home resulted in a U.S. Court of Appeals ruling that the sidewalk was actually public property, and the responsibility for keeping it clean could not be shifted to the owner of adjacent property.

1922 (September) – The city had purchased ten new snow plows and four new trucks. The 22 horse-drawn plows that had been used in the past were retired.

1925 – Motorized plows became an annual acquisition. A report on a storm in January 1925 mentioned the city's 33 snow plows, although they were deemed inadequate to the task of clearing 8 inches from the streets.

1920's – Chemical war on snow also began in the 1920s in many cities. Salt and calcium chloride (CaCl) replaced sand and ashes. Authorities in Washington had rejected the use of salt because of possible damage to pavement and to underground conduits for streetcars.

1931 – Morris Hacker, supervisor of city refuse, had 6 hydraulic loaders, 30 snowplows, 60 trucks and 30 horse carts to clear 17 miles and plow 120 miles of streets, 80 percent of which were in the business district. No plowing was done on residential streets. By comparison, the slightly smaller city of Hartford, Connecticut, used 18 snowplows and 4 mechanical loaders to clear 22 miles and plow 170 miles, 90 percent in the business district.

1930's – Washington struggled to keep pace with national trends in snow management. Parking bans from 2 to 8 a.m. were announced for some streets. The city promised to plow snow from 308 miles of main thoroughfares and boulevards and completely remove snow from 25 miles of streets in the business districts using 550 men, 250 trucks, 120 plows, and 15 mechanical loaders. The new snow management plans were tested February 7, 1936, when more than 14 inches blanketed the area. Predictably, although 4,000 men were hired to shovel, editors proclaimed snow removal inadequate, naming Congress for its failure to appropriate sufficient funds and city authorities for their "puny efforts."

1937 – The parking ban for snow removal was declared illegal by a District Police Court judge, who found it "arbitrary and unreasonable."

1940s – The war years saw an increased use of salt in the name of saving manpower and tires, since trucks had to spread only one layer of salt as opposed to three or four of sand.

1954 – As the city approached its greatest population, and its suburbs were also rapidly expanding, public discussion finally recognized the complexities of snow management. A four-inch snowfall in early January of that year caused schools and government offices to close.

1956 – The city agreed to pay the D. C. Transit

System for plowing and removing some of the city's snow, the contract included money for 15 trucks rigged with sand and salt spreaders in addition to the 80 already owned by the city.

1950's – Calgon Inc., of Pittsburgh, manufacturer of the polyphosphate Banox, and the National Aluminate Corporation of Chicago, maker of Nalco 8181-C, convinced hundreds of city governments to add their products to the salt spread on streets, claiming that the compound prevented corrosion to automobiles. Banox was artificially colored green, to signal that streets were protected "both against traffic mishaps and car damage," and to blend with the neatly trimmed and chemically treated lawns that symbolized the deseasonalized and oasis communities of "light snowfall cities."

1958 – After a 14 inch snowfall accumulation February 15-16, a new five-part Emergency Traffic Plan was put into operation, requesting the cancellation of all unnecessary travel, the prohibition of parking on principal streets, and the closing of federal and District offices. In the words of Superintendent of the Division of Sanitation William A. Xanten, "The entire metropolitan area is in for a severe case of "'stay put'."

1987 (January) – Mayor Marion Barry, in the tradition of East Coast politicians, was away during the storm of January 23. By ignoring the advice attributed to Mayor Richard Daley of Chicago-"just keep the streets clean and the buses moving and you can steal anything you want" – Barry joined Daley's successor Michael Bilandic, John Lindsay of New York, Stanley Makowski of Buffalo, Michael Dukakis of Massachusetts, and dozens of other politicians whose names are eponyms for snow storms of the 1960s and 1970s. Mayor Barry apologized for his administration's failure to clear the streets, but tried to excuse himself by observing: "I didn't bring it [the snow] here."

1987 (November) – Barry announced a new snow removal plan in Fall 1987 with an eight-page flyer explaining the parking ban and snow removal laws and new red and white street signs that read: SNOW EMERGENCY ROUTE NO PARKING DURING EMERGENCY. Barry's plan was tested sooner than anyone anticipated by the foot of snow that fell on Veterans Day

1987. Since none of the jurisdictions in the area was prepared, the District escaped blame for snarled traffic and school closings.

1993 – By the time the next heavy snowfall occurred, March 13, 1993, the city had a new mayor, and the Metropolitan Washington Council of Governments had instituted a plan that released government workers in four half-hour increments based on the distance they lived from the federal city. The *Washington Post* praised the city. Mayor Sharon Pratt Kelly was so pleased with her record on snow removal that she made it an issue in her re-election campaign. A flyer mailed to all District voters pictured a snowplow with the claim that her opponents – former Mayor Barry and Councilman John Ray – had incorrectly identified it as an ambulance. The absurdity of her ploy was reflected in her defeat at the polls.

1996 (January) – Nature joined politicians in shutting down the Federal Government for a month. The dispute between the President and Congress over the 1996 budget had just been resolved when the first of two storms struck. The storms, which brought between 17 and 24 inches of snow to the Washington area between the 7th and the 11th resulted in the usual "storms of protest" over street plowing. Again the District, with 2,821 lane miles (miles of road times the number of lanes) to clear lagged behind nearby counties in Maryland and Virginia with more than twice that number. Mayor Barry, back in office with diminished powers, kept a low profile, while the City Administrator Michael C. Rogers dealt with the press. It was estimated that the snow removal cost the District $3.6 million, Maryland $15.2 million, and Virginia $40 million in overtime pay, private contractors, and equipment.

Bernard Mergen is Professor of American Civilization at George Washington University and author of <u>Snow in America</u> (Smithsonian Institution Press, 1997) and "Slush Funds: A History of D.C. Snow Management," <u>Washington History</u> 8, 1, Spring/Summer 1996, 4-15, from which most of the above was excerpted. He is currently working on a book about American obsessions with weather.

Hundreds of Washingtonians have come to the reflecting pool to ice skate, circa 1918. The winter of 1917-1918 was one of Washington's coldest winters. *Library of Congress*

CHAPTER THREE

COLD WAVES

 ithout a doubt, when it comes to cold waves that have impacted the Washington area over the past century or so, the "Great Arctic Outbreak" of 1899 and the Great Cold Wave of 1912 are the benchmarks.

During the Great Arctic Outbreak, on February 11, 1899, the mercury plunged to -15°F, which remains the District's all-time record low temperature. In fact, the frigid air reached all the way to the Gulf of Mexico and South Florida. Ice floes were reported at the mouth of the Mississippi where the river flows into the Gulf of Mexico, and blizzard conditions were reported in many parts of northern Florida!

The Great Cold Wave of 1912 also set records that still stand today. The 1912 event took place over a six-week stretch from January 5 to February 16. The brutally cold intrusions caused temperatures to plummet well below zero, and in the process turned large sections of the Potomac River and Chesapeake Bay into solid sheets of ice. Some notable temperature extremes recorded during that span included -8°F at Washington; -19°F at Laurel; -21°F at Falls Church; and a truly frigid reading of -40°F at Oakland, in far-western Maryland. Oakland's reading of 40°F below zero remains Maryland's lowest temperature ever recorded.

More recently, the winters of the late 1970's, including the fabled "Bicentennial Winter", the winters of 1982-1985, and the winter of 1993-94 were characterized by repeated blasts of Arctic air that brought record low temperatures and wind chills well below zero to Washington and its suburbs.

Location, Location, Location!

One of the keys to determining why a particular region is susceptible to invasions of Arctic air masses relates directly to its geography – in particular, its latitude and elevation, as well as its proximity to mountains, hills, and major bodies of water. In Washington, which is located at roughly 39 degrees north latitude, the sun rises high enough above the horizon to temper the winter chill. In fact, if Arctic air masses didn't budge, winters in northern Virginia, Maryland, and the District of Columbia would be more tolerable, even in the middle of January. But unfortunately, these cold air masses do move, and like runaway trains, they're hard to stop once they get going.

Although a long way from Canada and Alaska, the Washington area is well within striking distance of the bone-chilling air masses that originate there. The lack of an east-west mountain range to hinder their southern migration is probably the most significant reason these air masses get so far from where they form. But, their movement over warmer ground as they head south, passage over the Great Lakes, and even

their movement over a small mountain range like the Appalachians help to temper these air masses before they reach the Washington area.

The "Arctic Warehouse"

The long, winter nights in the ice-and-snow-covered high latitudes of Canada and Alaska provide ideal conditions for the build-up of Arctic air. On calm, clear nights, solar energy that was absorbed by the Earth's surface during the day is allowed to rapidly escape into space. This is referred to as *radiational cooling.* In fact, in many Arctic locales the sun barely makes it above the horizon during winter, and therefore very little warmth is received from the sun. The result is that an air mass in contact with the ground becomes extremely cold. Temperatures in January and February routinely fall to 30 to 40 degrees below zero and lower in the Yukon and Northwest Territories of Canada. In fact, Canada's lowest temperature of -81.4°F was measured at Snag in the Yukon Territory near the border with Alaska. (The measurement was approximated by a pencil mark on the thermometer because the scale stopped at -80°F!)

These cold, dry air masses are designated either *cP (Continental Polar Air),* or *cA (Continental Arctic Air).* Continental Arctic air masses usually originate north of the Arctic Circle, while Continental Polar air masses form a little farther south. Driven by strong north-to-northwest winds, this frigid air can be directed southward over the vast prairies of central and southern Canada before being unleashed into the U.S.

Occasionally, a cross-polar fetch of bitterly cold air will slide out of Eastern Siberia into northern Alaska and Canada. As the massive dome of shallow Arctic air barrels southward into the lower 48, it sends temperatures plunging! These "Siberian Express" outbreaks dominated the winter of 1983-84. On Christmas Day 1983, 125 records fell across the central and eastern U.S., while in January 1985, high winds and single-digit temperatures forced the Presidential Inauguration indoors.

The Role of the Appalachians and Rockies

The north-south orientation of the mountains in North America means that Arctic air moving from Canada is not impeded on its trip south of the border. However, before it reaches the Washington area, the Arctic air must cross the Appalachian Mountains. The previous chapter discussed the role the Appalachians play in trapping Arctic air over the piedmont areas east of the high terrain. Ironically, for big cities along the I-95 corridor, like Washington and Baltimore, the Appalachians also provide a little protection from receiving the full brunt of an Arctic blast. The reason is that as the Arctic air descends the mountains, turbulent mixing and com-

pression of the air results in a slight warming of the air mass, as air molecules are pushed closer together. The effect is very similar to what happens when you use a bicycle tire pump – the outside of the pump becomes very warm as air is compressed and pumped into the tires. This is why cities like Indianapolis and Cincinnati, at roughly the same latitude as Washington but on the western side of the Appalachians, have recorded temperatures as low as -27°F and -25°F, respectively, while Washington's all-time record low temperature is considerably higher at -15°F.

Cold if by Land, Warm if by Sea

One of the Washington area's coldest mornings of the last 20 years occurred on January 17, 1982. In the days leading up to this record-breaking winter morning, a huge reservoir of Arctic air had been building over Canada with readings as low as -62.7°F in the Yukon Territory of northwest Canada. As the massive area of high pressure shifted into the U.S., Arctic air spilled

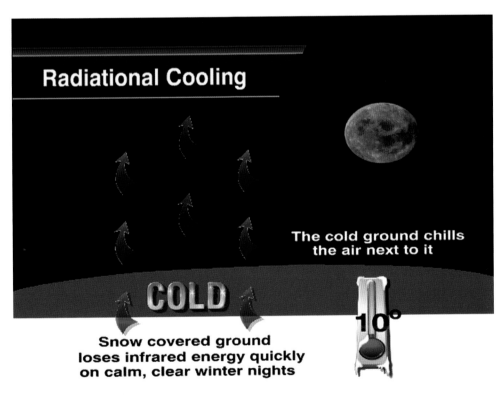

Radiational Cooling

The cold ground chills the air next to it

COLD

10°

Snow covered ground loses infrared energy quickly on calm, clear winter nights

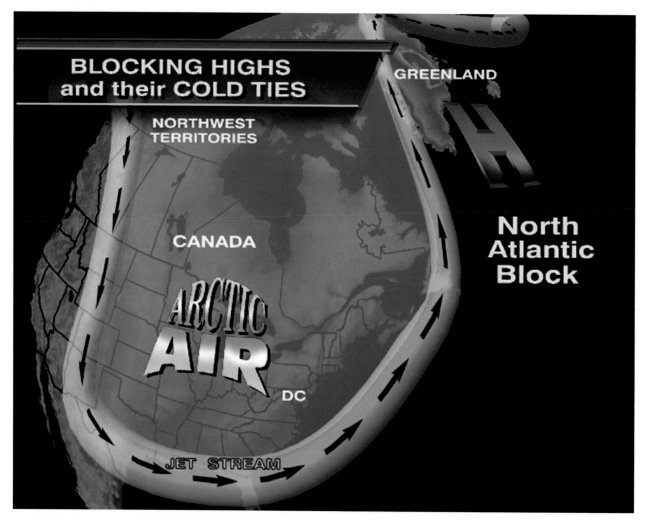

southward, sending the mercury plunging to -52°F at Tower, Minnesota and -45°F at International Falls, Minnesota. As the Arctic air mass pushed farther south, it moderated a bit. Temperatures "only" fell to 26 below zero in Milwaukee and 25 below zero at Chicago's O'Hare Airport.

By the time the Arctic express reached Washington, it had lost a bit more of its chilly bite. By all accounts, the reading of -5°F recorded at National Airport was truly a frigid morning for Washingtonians, but certainly nothing compared to what residents of the Upper Midwest had experienced with the same air mass.

While air masses do tend to lose some of their icy grip as they move south over warmer land, it's not nearly the same moderating influence cre-

ated by passage over hundreds of miles of 35 to 45°F ocean water. For example, cities like Seattle, Washington, and Portland, Oregon – both located at latitudes similar to northern Maine – are only exposed to Arctic air on rare occasions. Portland is occasionally hit by chilly blasts from the east that rip through the Columbia Gorge, but many parts of the Seattle metro area have never fallen below 0°F. The Pacific Northwest coast owes its moderate winter climate to its location with respect to the ocean and the mountains. Cold air masses are modified considerably as they cross the rather mild waters of the Pacific Ocean. These air masses are classified *Maritime Polar (mP)* and often bring cloudy, damp conditions to the Washington and Oregon coasts. Meanwhile, the towering Cascade Range and the Rockies to the east serve as nearly impenetrable barriers to arctic air surging down from western Canada.

The Washington area, on the other hand, does not enjoy the same luxury. The Appalachians are only one-third the height of the Cascades, and the Atlantic Ocean is located downwind from the metropolitan area. This means that strong arctic air masses usually have little difficulty passing over the mountains.

So, why does the nation's capital experience only a few intrusions of arctic air some winters, while other winters are beset with frequent arctic outbreaks? The answer lies in an examination of the large-scale flow of the atmosphere.

Blocking Highs and their Cold Ties

"Blocking Patterns" are characterized by large stationary high and low-pressure systems. These massive areas of high and low pressure act like boulders in a stream, forcing weather systems to slow down or take a detour around them. When a *"blocking high"* sets up over the North Pacific Ocean and another over the North Atlantic Ocean, a deep trough digs over eastern North America. The roller coaster wind pattern that develops, in which the upper-atmospheric winds dive toward the south into the *trough* and then bend sharply back to the north over the *ridge*, is called *"meridional flow."* With one high-pressure block located near the Gulf of Alaska, and extending south to the Pacific Northwest, and another block located near Greenland, mild Pacific air is prevented from entering the West Coast and bitterly cold arctic air is given a direct pipeline from the Arctic Circle into the con-

tinental U.S.

A blocking pattern may last for only a week before returning to a *zonal flow* pattern in which the upper-level winds blow predominately from west to east across the country. Zonal flow regimes usually spell rather mild weather for the Washington area as air masses originating over the Pacific Ocean move east and are warmed through compression as they descend the western U.S. mountain ranges and move onto the Great Plains on their way to the east coast. Typically, a zonal flow pattern will persist for a week before

January at noon in Resolute Bay in the Canadian Arctic. At the time this photograph was taken the temperature was –31°F. Cold air masses from Canada often move southward into the central and eastern U.S., bringing frigid temperatures and bitter windchills.
Vincent K. Chan

returning to a more amplified pattern featuring sharp troughs and ridges. This is why winters in Washington are often quite fickle. High temperatures may struggle to reach freezing for a week, followed the next week by a thaw in which temperatures climb well into the 40's and 50's.

Occasionally, blocking patterns will lock themselves in for weeks at a time. Such was the case during the harsh "Bicentennial Winter" of 1976-77, when the jet stream consistently dove all the way to the Gulf Coast, allowing arctic air to penetrate into South Florida. The longer a pattern like this holds, the colder each successive blast of arctic air is likely to be. As previously described, Arctic air masses are usually modified (warmed) as they move over the warmer land south of the U.S./Canadian border. The warmer ground heats the bottom of the air mass, making it unstable. This, in turn, causes convection currents that distribute the heat throughout the lower layers of the atmosphere. But, after a

winter's first few cold outbreaks, the land cools substantially and becomes much less effective as a "cold-air buffer." This is especially true if the ground is covered by a fresh snow pack. (Snow reduces temperatures because it reflects a great deal of incoming solar radiation during the day and is an excellent radiator of heat during the day and at night.)

During the Bicentennial Winter, records were shattered from Washington, D.C. to Florida. For the first time in recorded history, people in Miami, Florida witnessed snowflakes falling, while the Nation's Capital shivered through a prolonged cold spell. January's average temperature of 25.4°F made it Washington's coldest month since 1940.

New Wind Chill Index

		Temperature (°F)		
		30	20	10
		New Wind Chill Index		
		Old Wind Chill Index		
Wind (mph)	10	21	9	-4
		16	3	-9
	20	17	4	-9
		4	-11	-25
	30	15	1	-12
		-3	-18	-33

Some Chill Gone with the Wind

The *wind chill temperature* is a measure of relative discomfort due to the combined effects of cold and wind on exposed skin. As wind speeds increase, heat is carried away from the body at an accelerated rate. For years a debate raged in the scientific community concerning the accuracy of wind chill readings which were computed using an equation developed in 1945 during experiments in Antarctica. The outdated formula was based on the rate at which a can of water cooled as it was exposed to the wind.

Advances in science and technology have resulted in a new wind chill formula unveiled by the National Weather Service in November 2001. The new equation is based on experiments conducted in a chilled wind tunnel in Toronto, Canada, in which the faces of people were exposed to varying temperatures and wind speeds. Researchers found that the old formula appeared to overestimate how cold the wind makes it feel to the human body. Consequently, the new formula requires stronger winds or colder temperatures to match the old wind chill temperatures. For example, consider an air temperature of 20°F and a wind speed of 20 mph. Combining the effects of temperature and wind, the old equation produces a wind chill temperature of -11°F. The new equation yields a wind chill of 4 degrees above zero, or a difference of 15 degrees. For reference, note that scientists have determined that wind chill temperatures of -19°F and lower caused frostbite in 15 minutes or less.

Coping with the Cold

Subfreezing temperatures, and occasionally sub-zero chill, is a fact of life for Washingtonians in the winter. In order to cope with the cold, like many other facets of life, it's best to heed mom's advice: Dress warmly and wear a hat! Wearing a hat is extremely important because the body can lose up to 40% of its heat through the head. In addition, the most effective way to retain body heat is to wear layers of loose-fitting clothing. (Wool clothing is better than cotton in that it continues to insulate even when it's wet.) Covering all of the body's extremities, including the ears, face, and hands, is your best protection against *frostbite*, the freezing of body tissue. All these measures are important, since prolonged exposure to very cold weather, particularly when very low temperatures are accompanied by strong winds, can be life threatening.

As the speed of the wind increases, heat is carried away from the body at an accelerated rate and can eventually lead to *hypothermia*. *Hypothermia* occurs when the core body temperature drops below 95°F. Uncontrollable shivering, drowsiness, and disorientation are signs of hypothermia. If these symptoms occur, one should go inside immediately to get warm and seek immediate medical attention! Remember, children and the elderly are most susceptible to the effects of cold weather. Last but not least, pets should be kept in a warm, safe environment.

COLD WAVE EVENTS

The Late 1800's: The Coldest Temperatures on Record for Washington

At the end of December 1880, a deep snow cover of fourteen inches created ideal conditions for radiational cooling as an Arctic air mass settled into the region. The low temperature on December 31, 1880 fell to -13°F and the low temperature on January 1, 1881 was -14°F. Both readings set record low temperature marks for the months of December and January that still stand today. Thus, one air mass set the all-time cold temperature records for two months. The temperature readings were made at the Weather Bureau office at 24th and M Streets, NW. The Naval Observatory in Washington recorded -16.5°F on January 1, 1881.

Almost twenty years later, in February 1899, one of the coldest air masses on record invaded the southeast U.S. Washington's all-time record low temperature of -15°F was set on February 11, 1899. On February 11, many outlying areas recorded temperatures of -20°F to -25°F. Fredericksburg and Quantico recorded -21°F. The temperatures recorded at Washington during the 1899 cold wave were as follows:

February 9:   low of -4°F, high of 7°F
February 10: low of -8°F, high of 4°F
February 11: low of -15°F, high of 11°F
February 12: low of 4°F, high of 10°F
February 13: low of 4°F, high of 11°F

January 1912: Washington's Coldest Temperatures of the Twentieth Century

A moderate snowfall of 4 to 6 inches occurred on January 11-12, 1912. On January 13, the low temperature fell to -8°F and the high temperature recovered to only 8°F. The next morning, on January 14, the low temperature plummeted to -13. At College Park, Maryland, the temperature fell to an incredibly low reading of -26°F. January 1912 was the last month that

Clearing snow off of the ice, circa 1912. The photograph was taken on a pond or creek in the Washington area. *Library of Congress*

Ice blocks on the Potomac River push into Dempsey's Boathouse and the Washington Canoe Club, February 17, 1918. The moving ice, piled ten feet high, has crushed two buildings after the winter's first thaw created a huge ice breakup on the Potomac River. Record cold occurred during the winter of 1917-18, resulting in very thick ice on the river. During the thaw, temperatures warmed into the low 60's on February 12 and 15. This photograph was taken from the Aqueduct Bridge near Georgetown.

Library of Congress

A small house is taken down the Potomac River on an ice floe, February 17, 1918. Ten straight weeks of below-normal temperatures froze much of the Potomac River. A quick thaw in February caused the ice to break up, producing moving sheets of ice that overrode the shore-line, pushing this house into the river.

Library of Congress

A massive ice ridge has formed in the middle of the tidal Potomac River near Washington, February 1918. The interaction of currents, tides and wind has created this small mountain of ice on the river after the winter's first thaw. The Potomac River and Chesapeake Bay were locked in ice for most of January and early February 1918.
Library of Congress

double-digit sub-zero temperatures were recorded in Washington. Following the cold winter months of 1912, an extended period of mild weather began in September 1912, lasting 17 months.

Washington's Coldest Month: January 1918

Taking place during the middle of World War I, this record-setting cold winter caused severe hardships to our troops training in the field. In the Washington area, shortages of coal and kerosene developed. Many complained of problems with inadequately heated homes and offices. The Washington Post formed a coal committee and recruited a fleet of automobiles to deliver coal to thousands of needy families. Shipping was also hampered. The persistent cold froze the Chesapeake Bay to such an extent that powerful battleships were needed to open a narrow channel to Baltimore.

The winter of 1917-1918 occurred during a long period of colder than average weather. In Washington, every month from September through February had temperatures that averaged below normal. December 1917 is tied with 1989 as the second coldest December on record,

with an average temperature of 27.9°F. January 1918 is the all-time coldest month on record in Washington, with an average temperature of 23.6°F. (Overall, the coldest winter on record is the winter of 1904-05, which was 0.4°F colder than the winter of 1917-18).

The coldest period of the winter occurred from December 29 through January 4. During that week the temperatures at Washington were as follows:

December 29	High 17°F	Low 2°F
December 30	High 9°F	Low -3°F
December 31	High 12°F	Low 2°F
January 1	High 17°F	Low 1°F
January 2	High 13°F	Low 10°F
January 3	High 15°F	Low 4°F
January 4	High 20°F	Low 4°F

The snowfall for the 1917-18 winter tallied 36.4 inches. This is approximately twice the normal snowfall for Washington. The largest snowfall for the winter was 9 inches.

The Cold Wave of January 1977

The winter of 1976-1977 was one of the coldest winters in recent years. Washington's

EL NIÑO AND LA NIÑA WINTERS IN WASHINGTON

El Niño is a Spanish name for the Christ Child, and was first used by Peruvian fishermen to refer to the warming of the Pacific Ocean around Christmas. *La Niña* in Spanish means "the little girl" and, in many respects, is the opposite of *El Niño*, manifesting itself as the localized cooling of the Pacific waters.

El Niño is characterized by a weakening of the easterly trade winds over the equatorial Pacific Ocean, which allows warm water to surge eastward toward the coast of South America and block cold water from rising to the surface. La Niña is the reverse – east to west trade winds become stronger and warm water surges toward the western Pacific, while cold water from deep in the ocean rises to the surface along the South American coast. The trade winds help to push the colder water westward along the equator.

In the Pacific Ocean, huge clusters of thunderstorms follow the warm ocean waters – eastward during El Niño, and westward during La Niña. Winds in the upper atmosphere derive a lot of their energy from these thunderstorms. Consequently, as the warm water and thunderstorms shift direction, they change the jet stream winds and alter the course of larger-scale storms.

Specific to Washington-area winters, El Niño typically produces slightly warmer-than-normal temperatures and above-normal precipitation due to the increased occurrence of coastal storms. While about half of El Niño winters are rainy, the other half typically result in snowfall averaging around 150% of normal. During a strong El Niño event in 1983, Washington was buried by a massive snowstorm in February that dumped up to two feet of snow on the area.

La Niña, on the other hand, tends to produce above-average temperatures and below-average precipitation. That usually spells a mild, rather snow-free winter for Washington. Coincidentally, many of the snowiest winters in Washington have occurred in years without El Niño or La Niña.

Ships locked in ice on the frozen Potomac River, February 1936. The winter of 1935-36 was exceptionally cold and snowy. Ice conditions on the Potomac River and Chesapeake Bay were the worst since the winter of 1917-18. Steam-powered icebreakers were unable to keep all of the shipping channels open.
Washingtonian Division, D.C. Public Library

Icicles at East Potomac Park, February 15, 1946. Northwest winds gusting over 40 mph with an air temperature of 20°F created these curved icicles, as wind-whipped spray from the Potomac River froze on contact with the railing and bench. The strong winds and cold temperatures occurred after a cold front passage dropped temperatures nearly 50°F from a high temperature of 68°F on the previous day. *NOAA Library*

Top: Ice from the Potomac River over-rides the shoreline along Canal Road near Key Bridge, February 17, 1948. Cold weather in January and February 1948 caused thick ice to form on the Potomac. A warm-up in mid-February caused the ice to break up and move down river.
Copyright Washington Post; Reprinted by permission of the D.C. Public Library

Left: Keeping the ears warm on a brisk winter day, January 8, 1973. This photograph was taken at 13th and F Streets, NW. The high temperature was 24°F, and the low was 20°F.
Copyright Washington Post; Reprinted by permission of the D.C. Public Library

Left: Frozen water fountain, January 21, 1976. The winter of 1975-76 was very mild and virtually snow-free in Washington. Despite the mild winter, a few cold days still occurred. On this day, the high temperature was 33°F and the low was 26°F. *Copyright Washington Post; Reprinted by permission of the D.C. Public Library*

Bottom: Museum-goers dress warmly for the cold, January 29, 1977. The high temperature was 28°F, and the low was 8°F. Large crowds flocked to see the King Tut exhibit at the National Gallery and *"To Fly"* at the Air and Space Museum in 1977. January 1977 was Washington's fifth coldest month on record, with an average temperature of 25.4°F. *Copyright Washington Post; Reprinted by permission of the D.C. Public Library*

average temperature for January 1977 was 25.4°F. That was the coldest month in Washington since January 1940 (24.9°F) and not far from the coldest month on record, January 1918 (23.6°F). A thawing period occurred late in January when temperatures rose into the forties, from January 25 to January 28, which ensured that the 1918 record would not be broken. January 1977 was only the second month on record where the minimum temperature dropped below freezing every day of the month.

A cold weather pattern became established in December 1976 and it persisted throughout the entire month of January 1977. The month of January was one of the coldest months ever recorded in the Midwest U.S. The cold wave targeted the Ohio Valley, where stations smashed their all-time monthly records by several degrees and averaged up to 20°F below their monthly normal.

The cold reached its peak between January 16 and January 20, when the most powerful Arctic blast of the winter hit Washington. The coldest day of this frigid month occurred on January

18, when the minimum temperature of 2°F and the maximum temperature of 18°F set records for the date in Washington.

The same cold wave produced an unprecedented snowfall in central and southern Florida. For the first time in recorded history flurries were noted in Palm Beach, Miami and Homestead. Two inches of snow fell at Plant City, Florida (east of Tampa) and snow even fell at Freeport in the Bahamas!

Late in the month, headlines were made when Buffalo, New York was hit by its worst blizzard on record. In Buffalo, 70-mph winds whipped 30+ inches of snow into huge drifts, paralyzing the city for many days.

In Washington, three small snowfalls took place in rapid succession, between January 5 and

Ice skaters on the Potomac River near the 14th Street Bridge, January 1977. In January 1977, the low temperature in Washington was below freezing every day of the month. This is only the second month on record this happened in Washington.

Copyright Washington Post; Reprinted by permission of the D.C. Public Library

Ice on the Chesapeake Bay at the Bay Bridge, January 1977. Ice interfered with boat traffic and oyster harvests. Few Middle Atlantic winters are cold enough to freeze over the Chesapeake Bay. A cold weather pattern became established in December 1976 and persisted through January 1977, freezing both the Potomac River and Chesapeake Bay.

Copyright Washington Post; Reprinted by permission of the D.C. Public Library

January 10, covering the ground for the rest of the month. Little snow fell after that sequence. On January 14, an ice storm resulted in numerous accidents. On the Beltway, a tanker flipped and burst into flames near Tyson's Corner.

At the beginning of January 1977, ice had already formed on the Potomac River and Chesapeake Bay due to an unusually cold November and December. That situation worsened in January, as both bodies of water froze over. By January 20, Smith and Tangier Islands in the lower Chesapeake Bay could no longer be reached by barge, and food and other key supplies had to be airlifted to the residents. Ice cover on the Potomac River was up to 11 inches thick. In February, the situation became progressively better as temperatures returned to more normal levels and the ice thawed.

The Cold Waves of the 1980's and 1990's

Sub-zero degree temperatures occurred 28 times from 1870 to 1935. For 47 years, between 1935 and 1982, Washington had no sub-zero temperatures. Then, in the early-to-middle 1980's, a series of very cold, Arctic outbreaks produced two sub-zero temperature readings in Washington. (Note: Washington's cold weather statistics have been slightly skewed by the move of the city's weather station from 24th and M Streets to National Airport in 1942, and by the growing urbanization of Washington, which has created a "heat island effect.")

In January 1982, two severe cold waves hit in quick succession. On January 11 and January 12, the low temperature for both days was 2°F and the high temperatures were 9°F and 11°F, respectively. A week later, an even colder Arctic air mass moved into Washington with a low temperature of -5°F on January 17. This was the first sub-zero temperature reading since January 1935 and the coldest temperature since February 1934. Cold weather remained entrenched in D.C. for the next ten days. By January 31, temperatures warmed considerably and the Washington area recorded high temperatures in the low 60's.

On January 21, 1985, Ronald Reagan's second Inauguration was the coldest on record. The low temperature on January 21 was -4°F and the temperature at noon was only 7°F. Wind chill temperatures were well below zero. The oath of office was moved inside to the Capitol Rotunda and the parade was cancelled. The official high temperature for January 21, 1985 was 17°F.

During the middle 1990's, more severe Arctic cold hit Washington. On January 19, 1994, the low temperature fell to -4°F and the high temperature was only 8°F. The cold temperatures led to rolling power brownouts and closed schools and businesses. Ice glistened on trees for days in the prolonged cold that followed the numerous ice storms of January and February 1994.

With the repetitive ice storms and cold weather, ice cover on the ground was thick enough to support ice-skating on lawns. At the Penderbrook Golf Course in Fairfax, Virginia, downhill ice-skating occurred for several days on the long, sloped fairways. Farther north, heavy snow fell in New England, and Lake Superior froze over for the first time in 16 years.

In a swirling snowstorm, police and security inspect the podium where President Ronald Reagan was supposed to have been sworn into office for a second term, January 20, 1985. On the following day, extremely cold temperatures forced the swearing-in ceremonies to be moved indoors. Reagan took the oath of office in the Capitol Rotunda, and the Inaugural parade was cancelled. The low temperature on January 21 was –4°F, and by noon the temperature had only risen to 7°F.
AP/WIDE WORLD PHOTOS

Ice-covered Volkswagen in Columbia, Maryland, February 11, 1994. Frequent ice storms and very cold temperatures left the Washington area coated in ice for many days during the winter of 1994. *Ovidio DeJesús.*

An ice-covered Potomac River borders the Kennedy Center, January 14, 1996. The winter of 1995-1996 was cold and snowy in the Washington area. *Kevin Ambrose*

Cold weather has transformed the reflecting pool into an icy setting for the Washington Monument, January 14, 1996. Temperatures in January 1996 averaged below normal, with the coldest temperature of the month falling to 14°F.
Kevin Ambrose

Strong winds associated with a severe thunderstorm have uprooted this large tree along 17th Street near Constitution Avenue, July 18, 1971. The peak wind gust at National Airport was clocked at 48 mph, with 0.72 inches of rain.

CHAPTER FOUR

SEVERE WEATHER

erhaps the most awe-inspiring sight in nature is the ominous, black funnel of a strong *tornado*. By definition, a tornado is a rapidly rotating column of air in contact with the ground and pendant from a thunderstorm cloud. The thunderous roar associated with a tornado, often described like the sound of a freight train, can be heard from miles away and may be the only warning received by those caught helpless in its path. That distinct roar echoed throughout southern Maryland on April 28, 2002. On that day, one of the strongest tornadoes ever to hit the Washington area was spawned by a fast-moving *supercell* that tracked for over 200 miles, from West Virginia to near Ocean City, Maryland. A supercell is an intense, long-lived thunderstorm that possesses strong rotation. (Supercell thunderstorms are explained in more detail later in this chapter.)

An Atmosphere Ripe for Tornadoes

April 28, 2002, was a textbook setup for severe weather. The Washington area was entrenched in a very warm, humid air mass ahead of a strong low-pressure center. In addition, a vigorous upper-level trough associated with cold air and a belt of strong winds in the upper atmosphere was forecast to pinwheel across the Middle Atlantic region. This was expected to trigger an eruption of powerful thunderstorms, with the potential for supercells. Supercells often cause dangerous weather in the form of high winds and large hail, and occasionally produce intense tornadoes that track for many miles. Such was the case on this particular afternoon. The first tornado, rated an F2 on the Fujita Scale, touched down near Interstate 81 in Shenandoah County shortly before 5 p.m. and destroyed or damaged nearly 100 homes, farm buildings, and businesses. Although the supercell exhibited strong rotation as it sped across the northern Virginia countryside, the only reports of severe weather were in the form of golfball-size hail and high wind gusts that downed some trees and power lines.

As the severe thunderstorm approached the Potomac River, it encountered low-level southeasterly winds funneling up the Potomac. Prior to this, winds at the surface were blowing out of the southwest. It is believed this sudden shift in wind direction helped to increase rotation in the storm that led to formation of a second, much stronger tornado. This tornado, which touched down on the Maryland side of the Potomac River in Charles County around 7 p.m., blazed a 68-mile trail of destruction. Residents who were away from their television sets or simply unaware of the impending storm, had only seconds to react to the fast-moving twister that sped along at up to 55 mph.

The events that played out on this late April afternoon were eerily similar to those witnessed by thousands of awestruck commuters at the height of rush hour on September 24, 2001. On that day, two supercell thunderstorms produced 5 tornadoes in Virginia and in Maryland. The fifth and final tornado touched down in West Hyattsville shortly after 5 p.m. This tornadic whirl reached its peak intensity with winds close to 200 mph as it plowed through the University of Maryland campus in College Park. Tragically, two female students were killed when the car they were in was picked up and carried several hundred yards over an eight-story dormitory before landing in a clump of trees. This deadly tornado cut a 17-mile swath of destruction through west Hyattsville, College Park, Beltsville, and west Laurel before dissipating near Savage. Debris from the university campus was found up to 60 miles away in Harford County, northeast of Baltimore.

"Tornado Alley East?"

The 1990's and early 21ˢᵗ century have featured a rash of tornadoes in Maryland and Virginia. In fact, the annual tornado record for the greater Baltimore/Washington area has been broken three times in this time period. The busiest year in Maryland history occurred in 1995, when 24 tornadoes were reported. These numbers dwarf statistics compiled from 1950-2000 that show an average of 7 tornadoes hit Virginia and 5 tornadoes hit Maryland each year. While alarming, the upswing is probably not an indication that "Tornado Alley" is migrating eastward. First, weather events tend to display a cyclical nature. For instance, the 1920's were very busy tornado years around the Nation's Capital, and the 1950s and early 1960s saw a dramatic increase in tropical storms and hurricanes along the East Coast. Second, today's sophisticated

A tornado touches ground near the Washington Monument, September 24, 2001. This tornado originated near Fort Belvoir, Virginia then skipped through Franconia and Alexandria before moving into Washington. It alternated between F0 and F1 strength, with maximum winds between 50 and 100 mph. Minor damage to trees and homes occurred along its path. It ascended and remained aloft over much of D.C., but would later touch down in Maryland and reach F3 force as it moved through College Park. *Michael Shore*

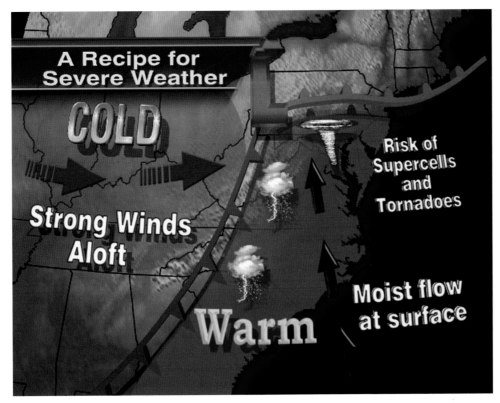

A Recipe for Severe Weather

COLD

Strong Winds Aloft

Warm

Risk of Supercells and Tornadoes

Moist flow at surface

troughs (wind-shift lines) that form on the east side of the Blue Ridge. These low-level boundaries provide a conducive environment in which air can spin and rise, and can help spawn tornadoes as they interact with approaching thunderstorms. The good news is that the overwhelming majority of these twisters are relatively weak and stay on the ground for just a few minutes.

However, severe weather in the form of destructive winds, large hail, and deadly lightning hits Maryland, northern Virginia, and the District of Columbia numerous times each year, particularly from April through July. One of the more destructive storms in recent memory occurred on April 23, 1999. Hail as large as softballs pummeled the landscape as a severe thunderstorm raced from the West Virginia panhandle southeast across northern Virginia. The storm smashed windows and windshields, dented cars and siding on homes, and shredded plants and trees. In total, the storm caused over $275 million in damage.

doppler radar technology has enabled meteorologists to spot tornadoes not visible in the early days. Finally, urban sprawl may also account for the increase in twister reports – development of previously uninhabited areas has resulted in more eyewitness accounts than ever before.

Mini Tornado Alleys around Washington

A county-by-county breakdown of tornado activity in the greater Washington area reveals something interesting. There appears to be a higher incidence of twisters in a corridor running east of the Blue Ridge through Loudoun and Frederick counties, and another "mini" tornado alley that stretches from St. Mary's County northward into Anne Arundel County along the western side of the Chesapeake Bay.

These two areas tend to be favored locations for low-level temperature and moisture boundaries, such as bay breeze fronts that propagate inland from the Chesapeake Bay, and lee-side

Thunderstorm Genesis

There are three basic ingredients needed to produce thunderstorms. First, moisture is needed in the low levels of the atmosphere to fuel the thunderstorm. Second, an *unstable* atmosphere is required to promote thunderstorm formation. The atmosphere is said to be unstable when temperatures aloft are much cooler than near the surface. Add moisture to the atmosphere and

Lightning over Potomac Heights, Maryland, June 2, 1998. Lightning is usually caused when negative charges in the base of a cloud are attracted to the positive charges at the earth's surface. *John Olexa, Jr.*

conditions are ripe for thunderstorms to pop. Third, something is needed to cause the air to begin rising. Air mass boundaries, such as *cold fronts* and *warm fronts,* are examples of such lifting mechanisms. (Cold fronts mark the leading edge of cooler air masses, while warm fronts denote the leading edge of warmer air masses.) Mountains also cause the air to rise by forcing it to move up their slopes. This process is called *orographic lift.* Finally, intense heating of the ground by the sun creates convective currents that force air to rise.

Sometimes a warm layer of air aloft will inhibit the lifting of air. In this case, the atmosphere is said to be "capped." It takes a cold front or strong jet stream winds to create enough lift to punch through the stable layer. When the lifting force is strong enough to "pry the lid open," clouds may soar to heights greater than ten miles.

Where There's Thunder, There's Lightning

Lightning is one of the most powerful and spectacular displays in nature. It occurs in all thunderstorms. Therefore, even a seemingly innocuous storm, which suddenly boils up on a

summer afternoon, is a potential killer.

So, how is lightning produced? The process begins when raindrops, and chunks of ice called *hail,* move up and down inside a cumulonimbus cloud (also known as a *thunderhead*), carrying negative and positive electric charges. Conditions for cloud-to-ground lightning are usually caused when negative charges in the base of a cloud are attracted to the positive charge at the Earth's surface. Positive charges may collect on buildings, trees, mountaintops, utility poles, and sometimes people.

Next, a *stepped leader* – a surge of negative electrical charge – descends from the cloud in zig-zagging steps, each step covering about 150 feet. When the stepped leader reaches to within 150 feet of a positive charge, a *streamer* or surge of positive electrical charge rises to meet it, and completes the electrical circuit. An electrical current then surges upward through the conductive path, and the path is illuminated from cloud to ground by a lightning stroke. The entire sequence occurs in less than two-tenths of a second. Many storms also produce reversed, positive cloud-to-ground lightning in which the flow of electrical charge is reversed. Positive lightning strikes tend to carry more charge and last much longer.

The temperature of a lightning bolt exceeds 50,000°F – about five times hotter than the surface of the Sun – which is why many lightning-strike victims suffer severe burns. It is this superheating which causes the air to expand and contract rapidly, creating sound waves which are heard as thunder. While cloud-to-ground strikes account for only about 20% of all lightning, they represent one of the most dangerous elements of a thunderstorm. In an average year, lightning claims the lives of about 90 people nationwide, and injures more than 400 others.

On June 17, 2000, a group of nine people were struck by lightning at a rugby match in Annapolis, Maryland, as they took shelter under a tree. One man was killed and two others were seriously injured. More people are killed by high-voltage current reaching out from a lightning bolt than by being struck directly. For this reason, high places and tall, isolated objects, which are prime targets for lightning, should always be avoided during thunderstorms. Lightning can also strike without a single drop of rain falling in the immediate area. Cloud-to-ground strikes have been documented to occur up to 10-15 miles from the parent thunderstorm. (This is referred to as a "bolt out of the blue.") A good rule of thumb to always remember is, "if you can hear thunder, you are close enough to be struck by lightning."

"What Goes Up, Must Come Down"

The upward moving air in a thunderstorm is called an *updraft*, while downward moving air is known as a *downdraft*. One of the keys to distinguishing a *severe* thunderstorm from the "garden variety" thunderstorm is the strength of the updrafts and downdrafts. By definition, a severe thunderstorm produces hail ¾ inch or larger and/or wind gusts of 50 knots (57 mph) or greater. Relatively weak updrafts and downdrafts

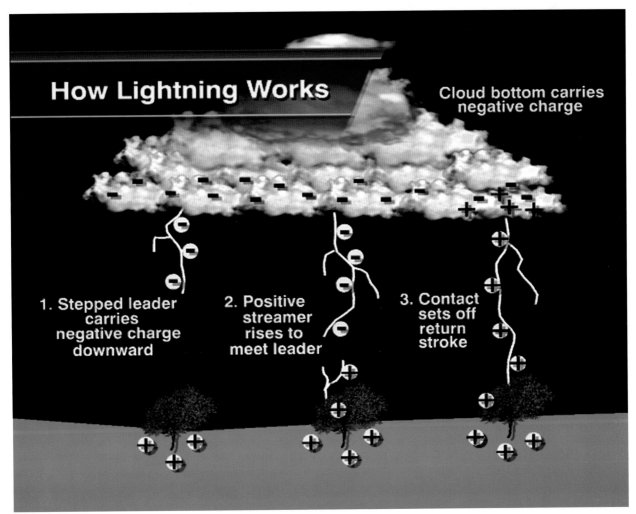

Shelf cloud in Sterling, Virginia. Shelf clouds signal the approach of a severe thunderstorm. This storm later moved into Montgomery County, Maryland where it caused considerable damage and one fatality. *National Weather Service Forecast Office/Sterling, Virginia*

are associated with non-severe thunderstorms. These *single-cell* storms may have a couple of updrafts and downdrafts that periodically strengthen and weaken. Also known as "pulse storms," these typically develop rapidly and may last only 20-30 minutes. However, in their short lifespan they are capable of producing torrential rain, hail as large as golfballs, weak tornadoes, and *downbursts*.

A downburst is a strong downdraft that includes a potentially damaging burst of winds near the ground. Localized downbursts, called *microbursts*, result as the downdraft hits the ground and spreads out in all directions, like water sprayed from a hose into the ground. If the storm is moving fast, most of the wind will blow in the direction of storm movement. Microbursts can cause damaging winds of 100 mph or higher, uproot large trees, blow down power lines, and destroy mobile homes.

Watch Out for "Bow Shapes" on Radar!

Multi-cell storms contain a series of updrafts and downdrafts that form in close proximity to one another. They can form as clusters or lines known as *squall lines*. Squall lines are often triggered as a strong cold front collides with a very warm, humid air mass. Often racing well ahead of the cold front, squall lines can be more than 100 miles long.

One of the first signs of an approaching squall line is a *gust front*, the leading edge of the thunderstorm-cooled air. Gusty, cool winds and sometimes a low-level, wedge-shaped cloud, called a *shelf cloud,* often mark such a squall line.

Squall lines are common culprits for producing severe weather such as high winds, dime-to-quarter-size hail, and flash flooding in the Washington area. If portions of the squall line start to "bow out," the potential for damaging winds becomes much higher. The bow or comma shape occurs when powerful winds slam into the squall line from the rear. This causes the middle of the squall line to accelerate faster than the rest of the line, forming a bow shape or *bow echo* on radar. Falling rain and hail from the storms help to drag the strong winds to the surface, causing downbursts that can reach speeds of 100 mph or greater. Bow echoes can be several miles long, last for hours, and occasionally produce tornadoes.

Supercells:
The Ultimate Thunderstorms

The major difference between thunderstorms that produce tornadoes and those that do not is the element of rotation. Unlike most thunderstorms that may contain several updrafts competing for moisture, a s*upercell* is highly organized. At the heart of the supercell is an intense, persistent, rotating updraft that ranges from about one to five miles in diameter, which is called a *mesocyclone.*

Wind shear plays a crucial role in these self-sustaining thunderstorms. (Wind shear is an abrupt change of wind speed or direction over a relatively short distance.) As winds aloft become much stronger, the updraft becomes tilted. This allows rain to fall outside of the area of rising air.

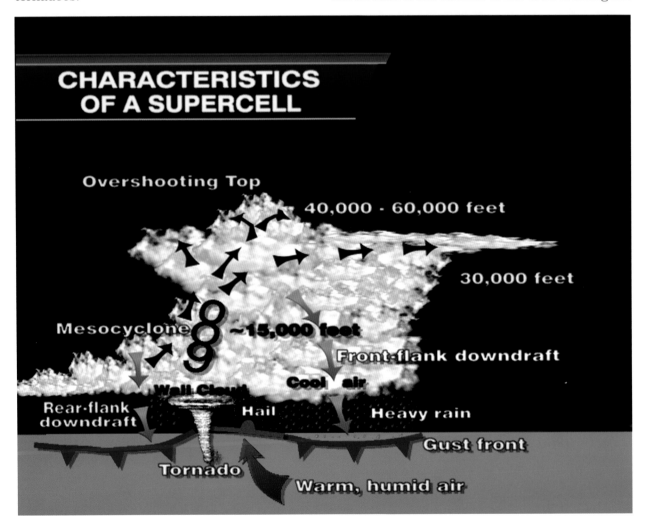

As a result, the rain-cooled air does not strangle the updraft and it can persist for hours.

The updraft's rotation also stems from wind shear. In a tornadic thunderstorm environment, winds generally blow from the southeast near the surface, and turn to the west with increasing height. Vertical wind shear can cause the air to spin much like a spiraling football. As the spiraling air enters the updraft, it is tilted upward, forming a mesocyclone around 15,000 feet aloft. Note: Dopplar radar

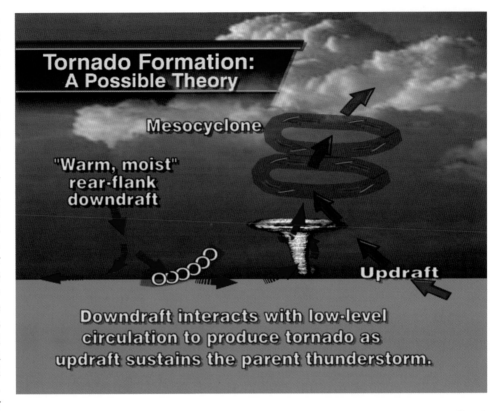

may indicate the presence of a mesocyclone by displaying a *hook echo*. This fishhook-shaped display is caused when precipitation is wrapped into the intense rotation. Hook echoes were detected on doppler radar for both the College Park and La Plata tornadoes.

Tornado Formation: Still Somewhat of a Mystery

Unfortunately, the process of how supercell thunderstorms transfer their rotation to a much smaller scale in the form of a tornado is not completely understood. One standard theory is based on what is called the *"dynamic pipe effect."* As air converges into smaller swirls within the mesocyclone, it spirals faster and faster like a spinning skater drawing her arms close to her body. As air gets sucked in from below, it circulates much faster, causing the rotating column of air to grow longer. This is a slow process, often taking 15 minutes or longer for the tornado to reach the ground. The latest research suggests

the interaction of a warm, moist *rear-flank downdraft*, descending at the rear of the storm with the mesocyclone, leads to the tornado (most downdrafts are associated with cool air). Due to its origination higher in the storm, the rear-flank downdraft contains a fair amount of rotation. As the rear-flank downdraft moves under the mesocyclone, it gets "sucked up" and tilted vertically. As the circulation is stretched, the air spins at much higher velocities.

In either case, it is believed the tornado dissipates as the rear-flank downdraft completely wraps around the circulation and cuts off the supply of warm, humid air. Occasionally, a new mesocyclone and tornado will form a few miles east of the dissipating tornado. These *cyclic supercells* have been known to produce several mesocyclones and tornadoes during their lifetime.

Research meteorologists have found that only about one-third of all mesocyclones produce tornadoes. In fact, the majority of twisters spawned, particularly in the Middle Atlantic region, are probably not associated with mesocyclones aloft. These nonsupercell

A powerful tornado moves through College Park , Maryland, September 24, 2001. The tornado was rated F3, with maximum winds reaching 175 to 200 mph. It cut a 17.5 mile-long path of destruction through Maryland and caused two fatalities. *WJLA and Megan Carpenter*

tornadoes, sometimes called *landspouts,* can form as a thunderstorm collides with a low-level boundary like a bay breeze front. Bay breezes, including those that blow in from the Chesapeake Bay, can often lead to low-level circulations. As the updraft from an approaching thunderstorm encounters this circulation, the circulation is tilted upward and stretched, resulting in the "spin-up" of a tornado.

Gustnadoes are another type of nonsupercell tornado that can develop along gust fronts. Research has shown that these "non-descending" tornadoes spin up very quickly, forming in five minutes or less. In some cases, these low-level circulations are either too weak or too far from the Doppler radar site to be detected. This is why trained, volunteer weather spotters will continue

to play a crucial role in providing severe weather data to the public for years to come.

Waterspouts: Tornadoes' Aquatic Cousins

A *waterspout* is a tornado that occurs over the ocean, a bay, or a large inland body of water. Many waterspouts begin as tornadoes over land and then move offshore, but a large number of them form over the water. These "aquatic cousins" of tornadoes are usually associated with very warm, relatively shallow water, and high humidity in the lowest several thousand feet of the atmosphere. For this reason, they are most common over the waters surrounding the Florida Keys, where 400 to 500 waterspouts are spotted each year. However, several sightings are reported annually on the Chesapeake Bay and over the Atlantic Ocean just off the coast of Maryland and Delaware.

First appearing as a dark spot on the water, a waterspout quickly develops into a dense, swirling ring of sea spray that climbs skyward. The funnel can rise to a height of several hundred feet or more, churning up the water as it moves along. Waterspouts are usually smaller, much less intense, and shorter-lived than their spiraling counterparts over land. However, some of the more energetic waterspouts can inflict quite a bit of damage if they make landfall. During the summer of 2000, a strong waterspout on the Miles River made landfall near St. Michaels, Maryland, uprooting several large trees, flipping over a boat, and damaging some small structures.

Tornado Danger Signs

Although there are mysteries still to be explained, much is known about tornadoes. For example, 80 percent of all tornadoes occur between noon and midnight, and are most likely to strike between the hours of 3 p.m. and 9 p.m. Also, the majority of twisters are relatively small – less than 50 yards wide – and remain on the ground for only a few minutes. Finally, seventy percent of all tornadoes, including the vast majority that hit the Washington area, are classified as "weak," packing winds of less than 112 mph.

Weak tornadoes are rated either F0 or F1 on the *Fujita Scale,* the scale of tornado intensity. The scale classifies tornadoes from F0, with winds of 40-72 mph, to F6, with winds of over 318 mph. (For all intents and purposes, the scale goes to F5 because an F6 tornado has never been observed) The most violent tornadoes, those with winds of over 200 mph, account for only 2% of all tornadoes. These devastating twisters may be over a mile wide, pack winds of 300 mph or even higher, and stay on the ground for tens of miles.

Fortunately, in many cases there are visual cues of the impending danger. These signs may include a dark, greenish sky, large hail (from golfball to softball size), or a nearly continuous display of vivid lightning.

Likewise, a rotating *wall cloud* definitely signals the need to take cover. A wall cloud is a cloud that lowers, often abruptly, from the

Waterspout on the Miles River near Saint Michael's, Maryland. It later made landfall as an F1 tornado, downing trees and flipping a boat. *National Weather Service Forecast Office/Sterling, Virginia*

base of the thunderstorm. Wall clouds that exhibit strong rotation and vertical motion often precede tornado formation by a few minutes to an hour.

Other tornadoes may occur with little or no warning – especially those concealed by a curtain of heavy rain or at night by a cloak of darkness. Also, not all tornadoes exhibit a visible funnel cloud. There may only be a dirt or debris cloud at the surface, while the funnel cloud remains several hundred feet above.

The Storm Prediction Center in Norman, Oklahoma – a branch of the National Weather Service – will issue a *Tornado Watch* when conditions are favorable for tornadoes to develop. The intent of a Watch is to give people as much advance notice as possible about the potential for tornadoes and/or severe thunderstorms. If a tornado is spotted or indicated by Doppler radar, the local National Weather Service office will issue a *Tornado Warning*. By the time a warning is issued, there may be very little time to react. The local National Weather Service Forecast Office in Sterling, Virginia issues tornado warnings for the greater Washington area.

Doppler Radar Detects a Near-F6 Tornado!

Doppler radar is an important severe-weather forecasting tool for a number of reasons. Doppler radar works by sending out microwaves, which are reflected by raindrops, ice crystals, hail, and even insects or dust. By measuring the frequency change of the reflected microwaves, the radar can determine if objects are moving toward or away from the radar site, and how fast they are moving. By pinpointing wind motions within the storm, meteorologists can detect mesocyclones, sometimes 20-30 minutes before a tornado develops.

Doppler radar, which is much more sensitive than conventional radar, can detect bound-

aries like gust fronts, low-level jet streams, and, of course, precipitation such as rain or hail. Storm chasers even use truck-mounted Doppler radars, called Doppler on Wheels (DOW) to look inside tornadoes. During a storm chase on May 3, 1999, scientists measured the fastest wind speed ever recorded, 318 mph, in one of the terrible tornadoes that hit the Oklahoma City area. (Note: This measurement was taken about 175 feet above the ground, and it is not certain whether wind speeds were quite as high at the surface.) The 318 mph wind speed put the tornado only 1 mph below the threshold for an F6 rating on the Fujita Scale.

Safeguarding Yourself and Others

Most tornado and severe thunderstorm deaths and injuries result from flying debris. Objects such as roof shingles, tree limbs, and glass shards become missiles that can pierce almost anything they strike. Obviously, this means that in almost all cases people are much safer inside than outside when severe weather strikes. A basement is the safest place to ride out a storm. If a basement is not available, take shelter away from exterior walls and windows, in an interior closet or bathroom on the lowest floor of the building. Because of their vulnerability to high winds, mobile homes should be evacuated for more substantial shelter.

If you are caught outside and there is no place to take cover, your only option may be to lie flat in a ditch and cover your head. Trying to outrun a tornado in a car often results in tragic consequences. In addition, contrary to popular belief, highway overpasses are not places of safe refuge during a tornado. Winds actually accelerate as they pass beneath the overpass. Instead, abandon the car and find a building to take refuge. If there is no substantial shelter to be found, lie flat in a ditch and cover your head. Common sense, advance planning, and quick reaction are the best defenses during times of severe weather.

Severe Weather Events

The Independence Day Storm of July 4, 1874

An intense line of thunderstorms swept into Washington from the northwest and provided a natural display of fireworks shortly after 8:00 p.m. on July 4, 1874. The same system had earlier drowned eight boaters on Lake Erie. As the dark, boiling clouds descended upon the District, intense winds hit, accompanied by blinding rain, brilliant lightning, and crashing thunder. From all quarters, there were reports of tumbling chimneys, flying roofs, and falling trees. In some instances, entire city blocks were totally unroofed. The Memorial Church, located at 14th Street and Vermont Avenue, and the Christ Church in Georgetown lost their roofs. Likewise, the downtown Masonic Temple and the Washington & Georgetown Railway Company lost their roofs.

Huge trees were reported "uprooted and strewn in all directions" in Lafayette Park and Franklin Square, across from the White House. Several historic trees on the Capitol Grounds also fell victim to the storm.

Although storm reports included "children flying through the air" with the storm's strong winds, no storm-related deaths were reported in Washington.

The Severe Thunderstorm of July 30, 1913

A typical, sweltering mid-summer heat wave held the Nation's Capital in its grip on July 30, 1913, as the mercury soared to 97°F. During the mid-afternoon, a thunderstorm crossed the center of Washington on an unusual northeast to southwest path, causing extensive damage. The storm's damaging effects were quite localized, confined to a path about four miles wide, covering downtown Washington and the nearby Virginia suburbs. No damage was reported in Maryland. While the press stated that the storm was tornadic, the damage apparently resulted from straight-line north winds. In addition, torrential rain and large hail buffeted the area.

The Weather Bureau office, located at 24th and M Streets, NW, measured the peak wind at 68 mph, with higher gusts recorded elsewhere. The storm dropped 2.10 inches of rain in thirty minutes. When the storm began, the temperature dropped nearly 30°F, almost instantaneously. The press compared the damage in downtown Washington to that of the Hurricane of 1896. Scores of homes were reported unroofed, while others were ripped off their foundations by the tremendous winds. Hail damage on the north side of buildings was widespread.

President Wilson was seated in the Executive Office when the terrible storm struck. Within seconds, three huge elms were uprooted and strewn across the lawn and several White House windows blew out. His aides escorted the President to an interior room as brilliant lightning and deafening thunder added terror to the scene. The White House grounds were described as "devastated."

At the Bureau of Engraving and Printing, thirty huge windows blew in and shattered on hundreds of employees. At the same time, several skylights also crashed down from the ceiling. Many workers were injured and sixteen were taken to Providence Hospital. In addition, 1000 new one-dollar bills blew out of the building during the storm. All but 30 bills were eventually recovered.

In the storm's most tragic episode, the three-story B. F. Saul Office Building, located at 7th & L Streets, NW, completely collapsed. Two employees, including a company vice president, were killed and several others were injured and trapped in the rubble.

The Silver Spring Tornado of April 5, 1923

During the daytime hours of April 5, a strong low-pressure system moved west of the mountains through Ohio, placing Washington in the unstable, warm sector of the storm. The dew points climbed into the low 60's, a high reading for early April in Washington. Sunshine boosted the temperatures above 70°F, which further destabilized the atmosphere.

In the early afternoon, a strong thunderstorm developed and moved northeast through Washington. The thunderstorm spawned a tornado which was first sighted in Rock Creek Park in Upper Northwest Washington around 2:30 p.m. Extensive damage to "many fine trees" was noted. The tornado then crossed the D.C./Maryland line and tore through Silver Spring, Maryland, passing about a quarter mile north of the Silver Spring railroad station. The path was 100-250 yards wide.

Witnesses described the sound of the approaching tornado as a rumbling sound, not unlike several large trucks on a rough road. Debris was sighted flying through the air. A distinct funnel-shaped cloud was clearly seen. Dozens of homes and buildings were destroyed.

The twister continued northeast for a total of 11 miles. It eventually crossed into Prince George's County near Avenel, Maryland, damaging additional buildings in that community. The twister was last seen near the south end of Laurel, Maryland.

Most of the damage in Silver Spring took place in a new subdivision, less than a year old. Seven homes were destroyed and a dozen more damaged. The Blair House was unroofed and approximately fifty trees on that property were downed. A baby was blown out of its house in Silver Spring, Maryland, but, miraculously, was unharmed and found on a nearby road. In a separate incident, two schoolgirls were picked up off of the ground and hurled through the air. They

also escaped injury.

A total of four injuries were attributed to the storm and property damage was estimated at $100,000. Tree damage in the uninhabited areas was described as incredible. On a side note, Washington had set its all-time record low temperature for the month of April only five days earlier, when the mercury plunged to 15°F on the morning of April 1.

Protect Your Property

TORNADO INSURANCE

We Represent Some of the Strongest Insurance Companies in America

Rates Cheerfully Quoted

RANDALL H. HAGNER & CO., INC.

1321 Conn. Ave.

Chas. E. Hagner, Mgr., Insurance Dept.

Tornado insurance advertisement in the Washington Evening Star, November 18, 1927. Three major tornadoes occurred in the Washington area during a five-year period, from 1923 to 1927. *Washingtoniana Division, D.C. Public Library*

The Deadly La Plata Schoolhouse Tornado of November 9, 1926

November can provide favorable conditions for severe weather in the Middle Atlantic region. Warm, humid air from the Atlantic can interact with winter-like storm systems, setting the stage for powerful thunderstorms. This was the situation on November 9, 1926, when a strong, low-pressure system moved just west of Washington, putting the D.C. area into a warm sector of air with temperatures above 70°F and dew-point temperatures above 60°F. These conditions helped to spawn a very strong and deadly tornado.

The disastrous tornado first touched down in Charles County, about 5 miles southwest of La Plata, Maryland. The tornado moved northeast into La Plata, then headed directly towards the local public grade school. One 11-year-old boy, seeing the tornado approach, jumped out of a school's window and escaped just in time. Seconds later, the entire schoolhouse, containing 60 students and two teachers, was lifted off its foundation and blown into a grove of trees about fifty feet away. The schoolhouse struck the trees and instantly splintered to bits, killing and injuring many of the children. Boards, furniture and children were thrown in all directions. Some of the

143

children were carried 500 feet, and one was found in the top of a tree over 300 feet away. A desk from the school was found five miles away from the scene, and some of the wreckage was found in Upper Marlboro, 25 miles away. A page from the school register was found in Bowie, 36 miles away.

Word of the disaster quickly reached Washington after an aviator from Bolling Field spotted the wreckage from the air. A massive rescue effort ensued as ambulances from the Army, Navy, and Red Cross raced to the scene from as far away as Walter Reed. The Marines were called in from Quantico, Virginia, to assist. The two teachers

Damage in Alexandria, Virginia from the Tornado of November 17, 1927. Trees, homes, and automobiles suffered significant damage in Alexandria. Ten storm-related injuries were reported.
Washingtoniana Division, D.C. Public Library

were bruised and battered but worked heroically with the rescuers to find the children.

Fourteen children, ages 7-12, lost their lives in this terrible disaster. Most of the other children were injured, some seriously. Twenty of the most severely injured children were rushed to Providence Hospital in Washington.

In the town of La Plata, five homes were entirely destroyed and four more severely damaged. Four tobacco barns were also demolished. Three additional people were killed in La Plata, bringing the final death toll for the tornado to 17.

The Alexandria and Washington Tornado of November 17, 1927

On November 17, 1927, a low-pressure system over western North Carolina moved north-northeast and passed 50 miles west of Washington. This put D.C. in the warm sector of the storm system. Numerous thunderstorms formed, with one thunderstorm in Fairfax County spawning a tornado. The tornado passed through western Alexandria and southeast Arlington, crossed the Potomac River at the Naval Air Station, and then passed through a densely-populated area of Washington near 19th Street and Bladensburg Road, NE. The tornado continued through Northeast Washington and crossed into Prince George's County. It was last seen in East Riverdale, Maryland.

The path of the tornado was 17 miles long and ranged in width from 20 to 300 yards. The tornado appeared most intense as it passed through Alexandria and again in Washington. Witnesses to the tornado reported that they saw a distinct funnel-shaped cloud and heard a roar, like a "huge waterfall." Debris was also seen flying through the air around the tornado.

Luckily, there were no fatalities, but 32 injuries were noted: 10 in Alexandria, 3 in Arlington, and 19 in D.C.

Tornado damage to row houses on Benning Road between 17th and 18th Streets, November 17, 1927. The tornado passed through western Alexandria and southeast Arlington, then through a densely populated area of Washington near 18th Street, NE. The path of the tornado was 17 miles long and ranged in width from 20 yards to 300 yards.
Washingtoniana Division, D.C. Public Library.

Tornado damage to houses on the south side of A Street, between 13th and 14th Streets NE, November 17, 1927. Most of these houses had their roofs sheared off by the tornado. There were 19 injuries reported in Washington.
Washingtoniana Division, D.C. Public Library.

A tornado seriously damaged this airplane at the Naval Air Station in Washington, November 17, 1927. The winds at the Naval Air Station were clocked at 93 mph as the tornado passed nearby. The wrecked airplane was reported to be a "bombing plane." *Washingtoniana Division, D.C. Public Library*

Wreckage of an automobile that had been lifted by the tornado and deposited on train tracks in Alexandria, Virginia, November 17, 1927. The ruins of a grain plant, destroyed by the tornado, are visible in the background.
Washingtoniana Division, D.C. Public Library

Lightning strikes the Washington Monument, June 7, 1937. The 555-foot marble and granite monument has been a frequent target of lightning strikes.

NOAA Library

Hailstones that fell in the yard on 13th Street NW, May 3, 1948.
Some of the hailstones that fell were noted to be the size of baseballs. Hail begins as ice crystals or frozen raindrops that are kept aloft by a thunderstorm's strong updraft winds. Hail often grows by repetitively making contact with supercooled water drops that freeze in layers on the hailstone's surface. Hail can grow larger than softballs.

Copyright Washington Post; Reprinted by permission of the D.C. Public Library

WASHINGTON'S HISTORIC LIGHTNING RODS

The Washington Monument's aluminum top and copper points have been blunted by 50 years of lightning strikes, November 19, 1934. At the completion of the Washington Monument in December 1884, a small and very expensive aluminum pyramid was placed atop the monument to function as a lightning rod. Aluminum was a very precious metal in 1884 and was chosen because of its white color and lack of tarnish. In less than a year, however, lightning had cracked the aluminum pyramid. Eight copper points were then added to the pyramid in 1885 to make it a better lightning rod. Despite looking like a "crown of thorns," the copper points were not visible from ground level. Over the years, the lightning protection system for the Washington Monument has been improved. The latest improvements occurred during the renovations of 2000. *Library of Congress*

Platinum-tipped lightning rod in the feathered headpiece of the Statue of Freedom, located atop the Capitol Dome, January 28, 1960. The Statue of Freedom was placed atop the Capitol Dome in 1863 and has ten lightning rods embedded in its headpiece, shoulders and shield. An ornate plate attached to the feathered headpiece reads, "These lightning conductors are manufactured and constructed by Carl Bajohr St. Louis, Missouri." Since the Statue of Freedom was put in place, platinum has increased in value and desirability, unlike the once-valuable aluminum used in the Washington Monument's lightning pyramid. The Capitol's lightning protection system is wired down from the dome into a stairwell. The Statue of Freedom was restored in 1993. *Library of Congress*

grade students about cyclones. When she saw the storm approach, she said, "Come to the window, children, this is a good example of our study." Suddenly, the door was ripped off its hinges and the teacher and several students were sucked out of the room and into the schoolyard. One child flew ten feet through the air and landed in a hedge. Fortunately, there were no serious injuries. The teacher, however, suffered a sprained ankle.

Later in the day, a waterspout was sighted over the Potomac River to the west of Anacostia. It moved to the southeast, but was short-lived and caused no damage.

The tornado passed over the Naval Air Station in Washington. There was a barometer and anemometer at the facility. As the tornado passed by, a very quick drop in air pressure from 29.57 to 29.11 inches of mercury occurred, with a peak wind of 93 mph from the south-southeast. At the Weather Service station, 3 miles to the northwest, no significant changes in wind or pressure were observed.

The tornado caused damage to hundreds of buildings in Alexandria and Washington. At the Naval Air Station, damage occurred to nine airplanes, fifteen automobiles and one destroyer. A huge smokestack was blown down at 12th and G Streets, SE, and a streetcar was blown off its tracks on Benning Road.

The most frightful event occurred at the Bryant School in Washington. A teacher had been lecturing to her fourth

The Police Patrol on flooded streets after the Thunderstorm of July 3, 1965. At National Airport, only 0.32 inches of rain fell, with a 30-mph wind. Several miles away, the rain was much heavier and the winds were much stronger. *Copyright Washington Post; Reprinted by permission of the D.C. Public Library*

The Severe Thunderstorm of July 3, 1965

On Saturday, July 3, 1965, a violent, fast-moving thunderstorm struck Washington with unusual ferocity. Although the storm lasted less than 15 minutes at any given location, it left an irregular path of destruction along its path from northwest to southeast across the metropolitan area.

The storm first struck in Montgomery and Fairfax Counties. In Clarksburg, Maryland, hail carpeted lawns like snow and stripped crops down to the stalk. In McLean and Arlington, Virginia, dozens of trees were blown down by strong winds.

Tree damage on the Mall, July 19, 1971. A severe thunderstorm hit downtown Washington the previous day, with 71-mph winds clocked at National Airport. *Copyright Washington Post; Reprinted by permission of the D.C. Public Library*

As the storm crossed into Washington, trees continued to topple. Hundreds of trees were blown down. Anacostia was the hardest hit, where tree damage was termed "as bad as Hazel." More than 20 District streets were completely blocked due to debris. Half a dozen cars in the city were reported crushed under fallen trees. Several minor injuries were also reported.

Extensive flooding on Wisconsin Avenue in Georgetown caused a large traffic jam. Likewise, a flash flood along Pimmit Run in Falls Church, Virginia, sent water into homes along the streambed.

The storm came while the Washington Senators were playing the Detroit Tigers in a baseball game at D.C. Stadium. The squall struck with such fury that it "threatened to wash D.C. Stadium down the Anacostia River." The game could not be resumed, and the Tigers prevailed 1-0 because the required 5th inning had been completed just before the storm hit.

The Mid-Winter Thunderstorms of January 26, 1971

A fast-moving cold front passed through the Washington area on January 26, 1971, triggering an unusual line of severe mid-winter thunderstorms. These storms were accompanied by rain, hail, snow and high winds, which reached 71 mph at National Airport and 78 mph in Silver Spring.

As a result of near hurricane-force winds, roofs were blown off several buildings and many trees were blown down. The wind also blew out windows at several office buildings in Montgomery County. Two injuries were reported.

A pilot spotted a tornado on the ground near Bethesda, but the National Weather Service never confirmed the tornado. In Laurel, golf-ball-size hail was reported.

Later in the year, on July 18, 1971, a severe thunderstorm struck downtown Washington causing extensive tree damage near the monuments.

The April Fools' Day Tornado of April 1, 1973

On April 1, 1973, a damaging tornado tore through Prince William County and Fairfax County in Virginia. The tornado was associated with a weather system that had previously produced deadly tornadoes in the Southeast U.S.

A car flipped by a tornado near Route 236 in Fairfax, April 2, 1973. The tornado first touched down near Manassas, Virginia then tracked to just north of Falls Church, Virginia. It did serious damage to the Pickett Shopping Center and Woodson High School in Fairfax, Virginia. *Copyright Washington Post; Reprinted by permission of the D.C. Public Library*

that caused eight fatalities in South Carolina and Georgia.

On the morning of Sunday, April 1, a surface low-pressure system was

Tornado damage to the Pine Spring Apartments near Merrifield, Virginia, April 2, 1973. A vivid example of why it is important to seek shelter on the lowest floor of a building during a tornado. In all, the tornado damaged 226 homes, with 53 homes declared uninhabitable. *Copyright Washington Post; Reprinted by permission of the D.C. Public Library*

over Illinois with a cold front cutting north to south through West Virginia. A warm front had passed through Washington, allowing unseasonably moist air to flow into the area. Then, as the cold front approached Washington, thunderstorms began to form.

A thunderstorm, moving northeast near Manassas, Virginia, spawned a tornado during the mid-afternoon. The twister skipped over a twenty-mile path to just north of Falls Church, Virginia. The first serious damage was noted south of Braddock Road at the Middle Ridge development, where a dozen homes were seriously damaged. The tornado then bounced back aloft before slamming into homes about a mile to the north, near the intersection of Braddock and Ox Roads. At least one house was lifted off its foundation and blown across Braddock Road.

The twister then hopped aloft again, next coming down about two miles to the northeast, near Little River Turnpike, where it did serious damage to the Pickett Shopping Center and Woodson High School. The tornado then made a final hop of three miles to the northeast before coming down in the Merrifield section at Lee Highway, near the Beltway. At that time, it sliced through the Pine Springs garden apartment complex and several other residences and businesses.

In all, 226 homes were damaged, 53 homes were declared uninhabitable, and 13 homes were totally destroyed. Also, 94 garden apartment units were damaged and 50 businesses sustained losses. Two elementary schools, Mantua and Oakview, were damaged. Woodson High School had its roof ripped off and several walls blown down. It was closed for the rest of the school year. Dozens of kids were in the Woodson gym playing basketball when the roof blew off. Luckily, there were no serious injuries.

In all, 37 injuries were reported as a result of the tornado. Fortunately, there were no fatalities. The total damage was estimated at $13.5 million.

The Crash of TWA Flight 514 in Severe Weather: December 1, 1974

Strong storms and high winds caused significant damage in the Washington area on December 1, 1974. The storms were believed to be a factor in the crash of TWA Flight 514 in Loudoun County, killing all 92 passengers.

TWA Flight 514 had originated in Indianapolis on Sunday morning, December 1, 1974, and was headed to National Airport. Due to a severe storm in progress in Washington, the flight was diverted to Dulles Airport. The plane disappeared from the Dulles radar as it was crossing the Blue Ridge Mountains. It was later learned the plane had crashed into the mountains at 1,750 feet above sea level, killing all 92 passengers on board. The speculation was that the plane had encountered turbulent conditions and high winds.

At the time of the plane crash, several waves of thunderstorms were moving through the area. Some observers reported as many as five separate thunderstorms. Widespread hail was reported in a ten-mile-wide swath, which ran roughly from Upper Marlboro to Rockville. Many observers reported the ground covered with hail along this path.

The winds peaked during the early afternoon, reaching a top speed of 61 mph at National Airport and 69 mph in Silver Spring. Downed trees and limbs blocked scores of area roadways. The Baltimore Washington Parkway was blocked as crews worked to remove fallen trees. Tens of thou-

sands of Washington area residents also lost power and phone service.

Rainfall amounts approached four inches, which caused widespread street flooding. A resident on 5th Street, NW, reported cars and a Metro bus floating up onto neighborhood lawns. In addition, the Potomac overflowed at high tide, flooding Maine Avenue, Hains Point and the Alexandria Waterfront.

In western Maryland and surrounding areas, the precipitation fell as heavy, wet snow. Snow depths reached 2 feet in the Frostburg area with up to three feet reported in Garrett County, Maryland and Somerset County, Pennsylvania. Heavy snow and high winds downed many trees and wires in the western mountains. Thousands of travelers returning from Thanksgiving vacation were stranded as they tried to drive through the mountains.

The Hailstorm of the Century, July 10, 1975

On July 10, 1975, an incredible hailstorm ripped through Loudoun County in Virginia and Montgomery County in Maryland. The hailstorm reached its greatest intensity near Poolesville, Maryland. For over 45 minutes, hail of golf ball size or larger pounded an area on both the Virginia and Maryland sides of the Potomac River. Winds

Hail covered ground near Poolesville, Maryland, July 10, 1975. Up to 5 inches of hail covered the ground near Poolesville and county snowplows were needed to clear some roadways. Later, the hail-chilled ground cooled surface temperatures and created dense fog that hampered cleanup efforts. *James Foster*

A forest near Poolesville, Maryland that was devastated by wind and hail from the Thunderstorm of July 10, 1975. Very strong winds and large hail stripped this forest. The area of severe hail damage was 10 miles long and 2 miles wide. *James Foster*

The Destructive Thunderstorm of July 16, 1976

An unusually strong mid-summer cold front triggered the development of a violent thunderstorm near Frederick, Maryland, during the mid-afternoon of July 16, 1976. The cell moved southeast at 30 mph through Rockville and Bethesda, then cut through northwest Washington reaching the downtown area around 5:00 p.m. The storm exited Washington in the Oxon Hill vicinity around 5:15 p.m. At any one location, the storm only lasted about fifteen minutes.

estimated at 85 mph added to the destruction. Approximately 2,000 customers lost power.

The area of maximum damage ran along a path approximately ten miles long and two to four miles wide. The hail fell with such force that some aluminum roofs were pierced, exterior siding was scarred and hundreds of windows were shattered. Vegetation and crops were also devastated. In Montgomery County, an estimated 6,500 acres of crops were destroyed.

At the storm's conclusion, up to five inches of hail covered the ground. It took many hours to melt the accumulation of hail. County snowplows were used to clear some roads. Hail filled depressions and stream basins up to three feet in depth. Despite temperatures in the 80's, hail deposits could still be found 72 hours after the storm in well-shaded areas.

Soon after the storm, the chilled ground temperatures and warmer air aloft caused extensive ground fog, which hampered cleanup efforts and gave an eerie look to an already devastated landscape.

A tree has crushed a car on 32nd and Cleveland Avenue NW, July 16, 1976. The driver walked away from the accident. Winds reached 76 mph at National Airport during the storm. *Copyright Washington Post; Reprinted by permission of the D.C. Public Library*

A car flattened by a tree that was blown over in a two-day windstorm, May 9, 1977. The car was parked at 31st and Q Streets, Northwest. Winds were from the northwest at 50 mph on May 9, and from the north at 42 mph on May 10. *Copyright Washington Post; Reprinted by permission of the D.C. Public Library*

Tornado damage at Williamsburg Square and Little River Turnpike in Fairfax, Virginia, September 6, 1979. The tornado was spawned by Tropical Storm David and severely damaged 22 homes in Fairfax. David spawned a total of 8 tornadoes along the Eastern Seaboard. *Copyright Washington Post; Reprinted by permission of the D.C. Public Library*

Trees downed by a severe thunderstorm in Chevy Chase, Maryland, June 14, 1989. Winds were clocked at 74 mph at Dulles Airport. Thousands of large trees in Chevy Chase, Bethesda and Washington were damaged by the storm. PEPCO reported that between 120,000 and 150,000 customers in Washington and Maryland lost power. *Rick Schwartz*

Many observers along the path of the storm reported very large hail and rainfall up to 2 inches. At National Airport, the peak wind reached 76 mph, the highest speed recorded since Hurricane Hazel. Tree damage was widespread. Along a path three miles long, between Wisconsin Avenue and the Potomac River, the tree fall was described as incredible, with 350 trees downed and 500 more severely damaged. The Park Service estimated that 100 trees fell on their property alone. On the White House grounds, the Adams Elm, an elm tree planted by President John Adams in the late 1700's, was severely damaged.

Most streets in Bethesda, Maryland and Northwest Washington were littered with fallen trees and branches. The scene was described "as if an army of lumberjacks had just come through." Considerable damage was also reported with trees falling on homes and autos. In a tragic incident on Reno Road, a woman was killed when a tree fell on her car.

Approximately 45,000 homes lost power in the metropolitan area. Traffic lights went out at 63 intersections and the D.C. police recorded 1,500 calls. In addition, several Washington radio and television stations were knocked off the air.

The Flag Day Tree Massacre, June 14, 1989

A strong front was stationary through Washington's northern suburbs during the afternoon of June 14, 1989. The front helped set the stage for a round of severe thunderstorms that downed thousands of trees in the Washington area.

At 2:00 p.m., the temperature at National

Cloud-to-cloud lightning lights the sky over the Mattawoman Creek in Charles County, Maryland.
John Olexa, Jr.

Geostationary Operational Environmental Satellite (GOES) image of a line of thunderstorms that produced a tornado in Fairfax County, May 6, 1996. The tornado moved east through Fairfax County, paralleling Route 29. Moderate tree damage occurred as the tornado skipped through Centreville and Fairfax, Virginia. *NOAA Library*

Airport was a humid 84°F while at Baltimore, Maryland the temperature was only 73°F. This temperature difference, combined with another front approaching from the west, helped fire up a line of severe thunderstorms that quickly moved through the D.C. area.

Shortly after 4:00 p.m., the squall line ripped through Dulles Airport, where a peak wind gust of 74 mph was clocked. The most severe storm crossed the Beltway near the Cabin John Bridge and continued almost due east across Bethesda, Chevy Chase and Takoma Park. The storm then weakened as it entered Prince George's County, but was still potent and caused significant damage in the Hyattsville, Maryland area. The path of the storm was over 25 miles long, but it was only a few miles wide.

Residential areas within the storm's path suffered damage to thousands of large trees. In many instances, the trees fell on wires, cars and houses. In one block of Bethesda, 200 felled trees were counted. PEPCO stated that this was the worst thunderstorm in its 93-year history. Between 120,000 and 150,000 customers lost power. The final damage estimate exceeded $50 million.

The Devastating Hailstorm of April 23, 1999

An isolated, severe thunderstorm barreled east-southeast from West Virginia and moved across the southern portion of the Washington area on the afternoon of April 23, 1999. The storm caused tremendous hail damage along a narrow path through the western and southern suburbs. The storm hit Hampshire County in West Virginia

Walking on hail stones in Burke, Virginia, after the Hailstorm of April 23, 1999. Severe hail damage occurred along a narrow path through western and southern Fairfax County. Hail causes $1 Billion in damage nationwide to property and crops each year. *John DiCarlo*

Hail on the ground is producing fog in Burke, Virginia, after the Hailstorm of April 23, 1999.
The fog formed when the hail cooled the surface air temperature to the dew point. The fog layer was quite shallow, but significantly reduced visibility on roadways.
John DiCarlo

and moved across Winchester, Virginia, between 3:00 and 4:00 p.m. The thunderstorm then moved southeast through central and southern Fairfax County where baseball-size hail was observed. The storm weakened with time, but still produced one-inch hail in Charles County, Maryland, shortly after 5:00 p.m.

The thunderstorm was a supercell storm that moved along a stationary front that was positioned across the southern suburbs of Washington. Afternoon temperatures just north of the front were in the 50's while temperatures just south of the front were in the upper 80's. The storm lasted about 20 minutes at any one location, but pro-

duced incredible hail damage to many towns and neighborhoods. The hardest hit areas were Chantilly, Centreville, Fairfax, and Springfield. Damage to roofs, cars, and siding was most often reported. Some neighborhoods suffered hail damage at almost every property. The ground along the storm's path was covered by up to an inch of hail.

Insurance adjusters were kept busy for over a year handling thousands of damage claims related to the storm. Signs appeared on street corners throughout Fairfax County advertising repair services for hail damage. One source reported that between 60,000 and 70,000 total claims were processed related to damage from the storm.

After the storm, the hail that had accumulated on the ground helped to create a dense, shallow layer of fog by cooling the surface temperatures to the dew point. The fog layer was about 10 to 20 feet thick and significantly decreased the visibility on roadways.

The Tornadoes of September 24, 2001

On the afternoon of September 24, 2001, two severe thunderstorms developed west and south of Washington and quickly moved to the north-northeast through the D.C. area. The thunderstorms spawned five tornadoes that began their destruction in Culpeper County, Virginia, and ended their deadly rampage in Howard County, Maryland.

The first tornado was associated with a severe thunderstorm in Culpeper County during the mid-afternoon. This tornado briefly reached F4 strength, with maximum sustained winds estimated between 200 to 225 mph. The most significant damage occurred in Rixeyville, Virginia, where a three-level brick home was blown to pieces. The tornado also produced damage in Jeffersonton, where three trailers were destroyed

and four churches were damaged. The path of the tornado was 10 miles long.

The second tornado was associated with the same thunderstorm and touched down near The Plains, Virginia, causing extensive tree and power line damage in Fauquier County. Wind speeds were estimated to be between 90 to 110 mph. The path of the tornado was 6 miles long.

Another severe thunderstorm developed in Spotsylvania County, near Fredericksburg, Virginia, shortly after 4:00 p.m. The thunderstorm moved north-northeast over Prince William, Fairfax and Arlington Counties. It then moved over D.C. and Maryland. This thunderstorm was responsible for producing three more tornadoes.

The first tornado skipped through Stafford and Prince William Counties. It was first seen aloft near Garrisonville, Virginia, and later touched down on the Quantico Marine Base. The twister was very weak, and only took down a few trees. It

Tornado moving through Laurel, Maryland, September 24, 2001. The tornado briefly reached F3 strength in Laurel, with maximum winds over 160 mph. Approximately 150 to 175 homes and businesses were damaged in Laurel, including Laurel High School. *WJLA*

A tornado and explosion (or lightning flash) silhouettes Byrd Stadium at the University of Maryland, September 24, 2001. The tornado, shown here at F3 strength, moved through the University of Maryland in College Park. Maximum winds were estimated between 175 to 200 mph. There were two fatalities and over 50 injuries associated with the tornado.

Dr. Ming-Ying Wei

A well-formed wall cloud and tornado near College Park, Maryland, September 24, 2001. The tornado remained on the ground for 17.5 miles as it moved north-northeast through College Park, Beltsville and Laurel, Maryland. This photograph was taken about six minutes after the photograph above. Both photographs were taken from an 18th floor balcony in College Park, about two miles from the tornado.

Dr. Ming-Ying Wei

Church bus blown into a grove of trees, Beltsville, Maryland during the Tornado of September 24, 2001. The bus had been parked at a church near Route 1 in Beltsville, Maryland. *Kay Grahm*

also passed through the Montclair community near Dumfries, Virginia, but again damage was minimal. The tornado was F0 strength, with maximum sustained winds of 50 to 70 mph. Its track was 11 miles long.

The second tornado touched down near Fort Belvoir, Virginia. This tornado skipped through Franconia and Alexandria before moving into D.C. It alternated between F0 and F1 strength, with maximum winds between 50 and 100 mph. The track of the tornado was 15 miles long. Minor damage to trees and homes occurred along the path. A neighborhood near Pentagon City was the hardest hit by the tornado, with large trees blown down and roof damage to several houses. The twister then passed by the Jefferson Memorial and the Capitol where it ascended into the sky.

The third tornado touched down two miles southwest of College Park and rapidly intensified to F3 strength. When it moved through the University of Maryland in College Park, it tossed cars, knocked down trees, damaged buildings and de-

stroyed ten trailers. The trailers were the temporary offices of the Maryland Fire and Rescue Institute. The northwest corner of the campus, not far from Byrd Stadium, received the worst damage. In all, twelve buildings at the University were damaged and approximately 300 cars were either damaged or destroyed.

The most tragic event of the storm took place when a car containing two women was lifted off the ground and hurled over an eight-story dorm building. The car fell into the woods across University Boulevard, killing both women. The women were sisters and students of the University of Maryland.

The storm struck particularly hard near the corner of University Boulevard and Metzerott

Debris was blown against this building in downtown Laurel, Maryland during the Tornado of September 24, 2001. The tornado briefly intensified to F3 strength in Laurel, with maximum winds over 160 mph. Serious damage occurred to the town's historic district. *Kay Grahm*

Tree damage on the University of Maryland campus after the Tornado of September 24, 2001. Some trees were sheared off near ground level. University Boulevard is seen in the background. *Andy Weiss*

THE BIRTH OF A THUNDERSTORM - AUGUST 2, 2002

1) Large cumulus clouds develop east of Washington and begin to move slowly south – 6:30 p.m.

2) A rainshaft (a column of falling rain) develops and expands just southeast of Washington – 7:15 p.m.

3) The storm grows and begins to produce lightning – 8:00 p.m. It remains almost stationary, just south of Washington.

Photos by Kevin Ambrose

4) A well-developed thunderstorm produces frequent cloud-to-ground lightning and heavy rain south of D.C. – 8:30 p.m. A flash flood warning was later issued for Alexandria, Virginia.

Road, where a church lost its steeple and sustained substantial structural damage. An adjacent apartment complex had its roofs torn apart. Near the University of Maryland Golf Course, the indoor tennis facility was completely swept away.

The tornado moved through the north side of College Park and then through Beltsville, tracking between Interstate 95 and Route 1. It was sustained at F2 strength, with maximum winds up to 150 mph. It felled thousands of trees and numerous power lines. The College Park Marketplace shopping center took a direct hit. Home Depot lost its roof and two other stores were rendered unusable. In addition, the roof of the St. Joseph's School in Beltsville was blown off into an adjacent building.

As the funnel churned into Laurel, it maintained its strength. It even briefly intensified to F3 strength. Approximately 150-175 homes and

businesses were damaged, including Laurel High School. There was also serious damage in the town's historic district.

The twister then culminated its rampage as it moved into Howard County, where it damaged 43 houses in the Settler's Landing community. In all, the tornado damaged or destroyed over 800 houses, 500 cars and 20 businesses along its 17.5-mile wide path through Prince George's and Howard Counties. There were two deaths and over 50 injuries. Damages exceeded $50 million.

The Devastating La Plata Tornado of April 28, 2002

The computer models for Sunday, April 28, 2002 revealed a potentially dangerous setup for severe weather. The atmospheric ingredients

The tornado that devastated La Plata, Maryland tracks through southern Calvert County, April 28, 2002. The tornado damaged or destroyed 860 homes and 194 businesses in southern Maryland.

A large tornado spins on the Chesapeake Bay southeast of Long Beach, St. Leonard, Maryland, April 28, 2002.
The tornado, associated with the supercell thunderstorm that hit La Plata, Maryland, was estimated to be two miles from shoreline at the time of this photograph. *Ted L. Dutcher*

Twin tornadoes race across the Chesapeake Bay away from Long Beach, St. Leonard, Maryland and towards Taylors Island, Maryland, April 28, 2002. This photograph was taken facing due east. *Gail Siegel*

included a moist, southerly flow at the surface ahead of a strong low-pressure center moving through northwestern Pennsylvania, a belt of strong winds aloft crossing the Middle Atlantic region, and an approaching cold front. As morning rain showers gave way to afternoon sunshine, the volatile air mass was heated, which added further instability to the atmosphere.

During the morning of April 28, the Storm Prediction Center (SPC) in Norman, Oklahoma forecasted a moderate risk of severe weather, including the risk of tornadoes, for much of Pennsylvania, Maryland, Northern Virginia, and Washington, D.C. As thunderstorms moved through West Virginia, the SPC issued a Tornado Watch for the entire Washington area. Soon thereafter, a supercell thunderstorm developed over eastern West Virginia. It spawned the first tornado at 4:45 p.m. in Shenandoah County, Virginia, and caused major damage. Over two dozen homes and farm buildings were demolished,

and 75 other homes, businesses, and farm structures were damaged.

As the rotating thunderstorm raced east through north central Virginia at 40-50 mph, it produced hail and hurricane-force wind gusts in Culpeper and Fauquier counties, with golf ball-size hail reported near Dale City, Virginia. When the supercell thunderstorm crossed the Potomac River around 7 p.m., a second, stronger tornado touched down along the southwest flank of the storm in Charles County.

The tornado stayed on the ground for nearly 70 miles as it sped along at 45-55 mph through Charles and Calvert Counties in Maryland. It then crossed the Chesapeake Bay and continued its rampage through much of Dorchester County

Top: An aerial view of the supercell thunderstorm that produced the La Plata tornado, April 28, 2002. The bulging dome of clouds extending above the supercell's flat, anvil top (called an overshooting top) is caused by a very intense updraft. The updraft is so strong that it has enough momentum to punch through the tropopause and into the stratosphere (the earth's upper atmosphere). *Steven Maciejewski*

Left: Lightning illuminates the tornado funnel in Calvert County, April 28, 2002. The tornado stayed on the ground for nearly 70 miles as it sped through Charles and Calvert Counties in Maryland.

Top: The path of the La Plata tornado is clearly visible from this satellite image, April 28, 2002. Trees, vegetation, and buildings have been destroyed or disturbed along the immediate path. *NOAA and NASA*

Right: A hook echo appears on WJLA's doppler radar, indicating the likelihood of a tornado, April 28, 2002. The fishhook shape is caused as precipitation is wrapped in the intense rotation. Dopplar radar, first used for meteorological measurements in the 1950's, has become a widely used tool in forecasting and analyzing the weather. *WJLA*

on the Maryland Eastern Shore. Observers reported seeing twin tornadoes near the Calvert Cliffs Nuclear Power Plant and over the Chesapeake Bay. In addition, baseball- and softball-sized hail was observed in the vicinity of Pomfret, La Plata, and Hughesville, Maryland. (The supercell thunderstorm that produced the tornadoes and hail in Maryland tracked from West Virginia to the Atlantic Ocean!)

One of the hardest hit areas was concentrated around the town of La Plata in Charles County. Parts of the quiet, southern Maryland community could only be described as a war zone. There was massive destruction in the downtown section, including the town's shopping center and business establishments – located adjacent to the intersection of Routes 6 and 301. Winds were so violent that some homes were completely swept off their foundations and trees were stripped of their bark. La Plata bank receipts were found 70 miles away by a man in Seaford, Delaware.

The storm and tornado damaged or destroyed 860 homes and 194 businesses in southern Maryland. Five lives were lost and at least 120 were injured. Property damage was estimated in excess of $100 million.

Officials from the National Weather Service said the tornado's winds fluctuated from 100 mph (F1 on the Fujita Scale) to nearly 260 mph (F4 on the Fujita Scale) during its 90-minute life cycle. Tragically, the tornado peaked to F4 intensity as it moved through the town of La Plata. The tornado was F4 strength for only one minute while it moved through downtown La Plata. Note: The National Weather Service initially rated the tornado an F5 on the Fujita Scale. However, subsequent surveys of the damage by structural engineers and meteorologists revealed the destruction was more consistent with damage caused by an F4 tornado, with winds of 207 to 260 mph.

Hail covers the ground in Pomfret, Maryland, April 28, 2002. *Eric Beach*

Baseball-sized hail that fell near Hughesville, Maryland, April 28, 2002. Hail this large will reach speeds of over 100 mph as it falls to the ground. *NOAA photo contributor*

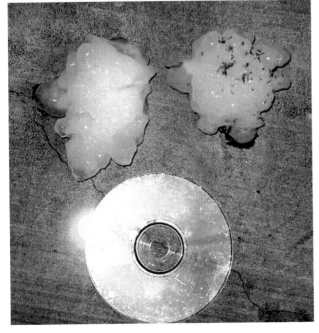

Hail stones approach the size of a CD-ROM in Hughesville, Maryland, April 28, 2002. The same supercell thunderstorm that produced the La Plata tornado also produced these large hail stones. *James T. Bailey, Jr.*

A helicopter surveys the damage in La Plata caused by the Tornado of April 28, 2002. This photograph was taken less than an hour after the tornado devastated the southern Maryland town. *John Olexa, Jr.*

Tornado damage to the La Plata shopping center, La Plata, Maryland, April 28, 2002. The tornado moved through town at 45 to 55 mph. *John Olexa, Jr.*

The KFC in La Plata, Maryland has been demolished by the Tornado of April 28, 2002. The tornado reached F4 status in La Plata, with maximum winds of 207 to 260 mph. *Andy Weiss*

A house near La Plata that was leveled by the powerful Tornado of April 28, 2002. Five lives were lost and at least 120 were injured. *NOAA*

Fighting the winds of Hurricane Hazel at East Potomac Park, October 15, 1954. Hurricane Hazel made landfall north of Myrtle Beach, South Carolina as a category 4 storm with winds of 135 mph. The hurricane then moved quickly north, tracking through the western suburbs of Washington. Hazel produced National Airport's all-time record wind gust of 98 mph. *Copyright Washington Post; Reprinted by permission of the D.C. Public Library*

CHAPTER FIVE

TROPICAL WEATHER

o other force in nature combines the sheer size and power of a mature hurricane. Beginning as a cluster of thunderstorms over warm seawater, hurricanes can grow into spiraling masses up to 600 miles wide, and pack winds of over 150 mph. While coastal residents bear the brunt of the raging winds and powerful *storm surge* (a hurricane-induced rise and onshore surge of seawater) the Washington area is by no means immune from the impacts of hurricanes and tropical storms striking the Gulf Coast and East Coast of the U.S. Many long-time Washingtonians still have vivid memories of storms like Hazel and Agnes that pounded the area with hurricane-force winds, and torrential rain that led to major flooding. A storm packing winds like Hazel and rain like Agnes could be devastating to the Washington area. Consider the Raleigh-Durham-Chapel Hill area, which is located well over 100 miles from the coast, yet suffered a billion dollars in damage in the wake of Hurricane Fran in 1996. During the onslaught, the Triangle region was hit by 60-80 mph winds and up to 10 inches of rain that toppled huge trees by the thousands on homes, businesses, cars, and power lines. Although a storm of this magnitude would be rare in Washington, it is possible.

Steering a Hurricane

The Washington area is most vulnerable to hurricanes that strike the Middle Atlantic coast, in particular from roughly Myrtle Beach, South Carolina, northward to Delaware. As they chug across the Atlantic Ocean, these hurricanes typically steer around the outer edge of the *Bermuda High*. This semi-permanent area of high-pressure dominates the flow across the western Atlantic during the summer months. When the Bermuda High builds westward toward the East Coast of the U.S., tropical systems generally pose a much greater threat of making landfall. Whether these storms strike the coast depends to a large extent on the upper atmospheric winds they encounter as they approach the eastern seaboard. The majority of these tropical cyclones are deflected to the north and then to the northeast before making landfall as they encounter the prevailing westerly winds that blow at high altitudes over the U.S. However, when the steering winds in the upper atmosphere are weak, these creatures of the sea are quite unpredictable. In 1999, Hurricane Dennis stalled, wobbled, and weaved along the East Coast for 10 days before finally coming ashore in North Carolina.

Fortunately, the hurricane threat – particularly the risk of an intense hurricane – decreases rather markedly along the Atlantic Coast north of Cape Hatteras, North Carolina. While the remnants of tropical systems impact some portion of Maryland or Virginia almost every year, only five hurricanes made landfall along the Virginia, Maryland, and Delaware coasts from 1900-2000. Of these five hurricanes, only the Chesapeake-Potomac Hurricane of 1933 packed winds greater than 100 mph when it made landfall

in southeastern Virginia. The relative scarcity of big, landfalling storms north of the Outer Banks of North Carolina is due in large part to eastward moving cold fronts and troughs that deflect them out to sea, and to colder water temperatures storms encounter as they pass north of the Gulf Stream. The cooler Atlantic waters cause hurricanes to weaken and lose their tropical characteristics. To understand why hurricanes weaken in the absence of very warm water, it is necessary to know how they operate.

Hurricane: The Atmospheric Engine

Hurricanes work much like atmospheric engines which are fueled by warm ocean water (at least 80°F.) As ocean moisture is evaporated, a rising plume of water vapor condenses into water droplets that become visible as clouds. During this change of state, called *condensation*, energy is released in the form of *latent heat*. As the air is heated, it becomes lighter. The buoyant air rises around the central core of the system, and then flows outward at high altitudes. The high-level outflow can be thought of as the storm's exhaust system. As air flows out of the top of the hurricane, air is pulled into the hurricane at low levels, spinning ever faster as it spirals into the center of the storm. As winds increase, more water vapor is drawn from the warm sea releasing more heat and driving the central pressure lower.

At the center of a hurricane is a relatively calm, sometimes cloudless area called the *eye*. Many crewmembers that have flown aboard the Hurricane Hunter aircraft have observed birds flying in the eye during the day, and have witnessed brilliant starlit skies at night. The most intense weather, including very strong, turbulent winds and blinding sheets of rain, occurs in an area bordering the eye called the *eye wall*.

Hurricanes in their infant stage are merely large clusters of thunderstorms. When the thunderstorm complex develops a closed circulation, it is classified a *tropical depression*. Once its winds reach 39 mph, it is classified a *tropical storm*. The storm reaches *hurricane* status when sustained winds reach 74 mph.

When a hurricane makes landfall, it is cut off from its energy source – the warm ocean water. Also, the increased friction the land imposes on the storm circulation causes the sustained winds to weaken. This is why systems that make it as far inland as Washington often arrive in a weakened state, as either a tropical storm or a tropical depression. Unfortunately, a storm named Hazel was anything but weak when she paid a visit to the nation's capital nearly 50 years ago.

Hazel Blows Through D.C.

In October 1954, Hurricane Hazel plowed ashore just north of Myrtle Beach, South Carolina. Hazel was an example of a strong, fast-moving storm that brought hurricane-force winds to locations hundreds of miles from where she made landfall. Hazel rated a strong Category 4 on the *Saffir Simpson Scale*. The Saffir Simpson Scale classifies hurricanes on a scale of 1 to 5, based on wind speed and the extent of damage done by

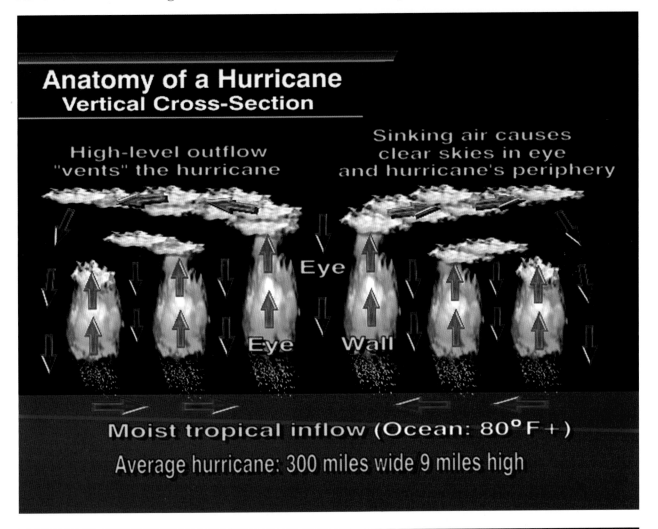

the storm. Wind speeds start at 74 mph for a Category 1 storm, and escalate to 155 mph or greater for a Category 5 rating. Hazel was still packing sustained winds of 80 mph, with gusts to 98 mph, as she steamed through D.C. In fact, the eye of Hazel was still discernible as it passed over Toronto, Canada! Like Hazel, Hurricane Hugo in 1989 was another example of a hurricane that left its mark on communities far removed from the coast. Located about 175 miles inland, Charlotte, North Carolina experienced wind gusts to 100 mph as Hugo roared through the southern piedmont of the Tarheel State.

Prompted by Hugo, researchers at the Tropical Prediction Center in Florida stepped up their work to develop a model to estimate the maximum sustained surface winds as a hurricane moves inland. The product of that research, "The Inland Wind Model," indeed showed that fast-moving major hurricanes have a higher potential for carrying strong winds over greater distances as they move inland. For example, a slow-moving (15 knots/17 mph or less) Category 1 hurricane (at landfall) has the potential to bring 35-50 knot (40-58 mph) winds to the Washington area,

especially to places like Anne Arundel, St. Mary's, and Calvert Counties, which are much closer to the Chesapeake Bay. The threat of even stronger winds reaching the Washington area becomes much greater for a Category 3 storm (at landfall) moving at a forward speed of 25 knots (approximately 30 mph). In this case, hurricane-force winds (65-80 knots/74-92 mph) envelop a good portion of the metro area, particularly from Washington east to the bay.

The Inland Wind Model's calculations are based on the scenario of a storm moving inland perpendicular to the coastline. The strongest winds are encountered on the right side (northeast quadrant) of the storm, where the storm's forward motion contributes to the wind speed. For sustained, hurricane force winds to be felt in the Washington area, the storm would have to make landfall near Rehoboth Beach, Delaware, or perhaps Ocean City, Maryland, and move almost due west across the bay. This is a very unlikely storm track since most tropical storms near D.C. have a northerly component to their movement. A more likely scenario would involve a Category 3 hurricane making landfall in southeastern Virginia or the northern Outer Banks, then moving north-northwest along the Chesapeake Bay. Moving over the warm bay water would likely allow the storm to sustain its strong wind field as it moved inland, particularly if it increased its forward speed. Illustratively, the Chesapeake-Potomac Hurricane of 1933 followed a very similar path from Nags Head, North Carolina, through Norfolk, then paralleled the Chesapeake Bay on

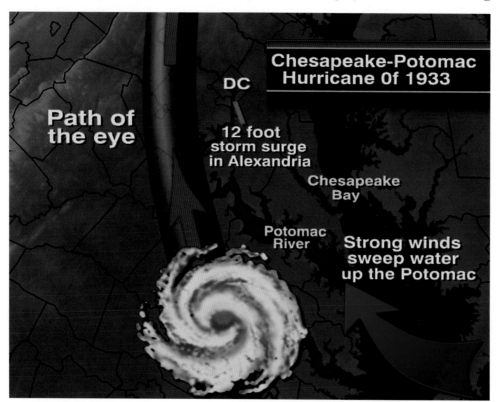

Chesapeake-Potomac Hurricane Of 1933

Path of the eye

DC

12 foot storm surge in Alexandria

Chesapeake Bay

Potomac River

Strong winds sweep water up the Potomac

its way toward Washington. The rather slow movement of the storm contributed to its weakening over land. Consequently it produced a maximum wind gust of only 58 mph in Washington during the passage of this storm, but its track just west of the bay resulted in unprecedented flooding in places like Alexandria and the District of Columbia.

Storm Surge on the Potomac and the Chesapeake Bay

In most cases, tidal flooding on the Potomac River that results from the approach of a tropical system usually produces "nuisance flooding." Tides with storms like Fran and Floyd ran about three to five feet above normal and caused only minor street flooding. However, communities on the Chesapeake Bay are much more vulnerable to a storm surge. Eastern Anne Arundel County, Kent and Tilghman Islands, St. George Island in St. Mary's County, and most of Dor-

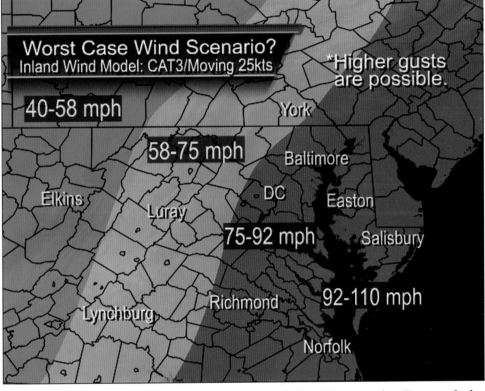

chester County (south of Rt. 50) on the lower Maryland Eastern Shore would suffer minor-to-moderate flooding by a Category 1 hurricane moving up the bay. But in cases of major storms taking an inland track, flooding becomes much more pronounced.

The Chesapeake-Potomac Hurricane of 1933 will long be remembered for its devastating hurricane tide. The storm hit as a strong area of high-pressure was building over New England. The two forces working in tandem produced a long, east-

erly fetch of wind that piled up water along the Atlantic coast. After making landfall in Nags Head, North Carolina, the storm moved directly over Norfolk and then marched just west of the Chesapeake Bay toward the nation's capital. As winds shifted to the south, a Chesapeake Bay *tidal bore* began to develop. A tidal bore is a solitary wave, formed when a rising tide enters a shallow, gently sloping, and narrowing river from a broad estuary. Because the hurricane traveled at about the same rate of speed as the tidal bore, it continued to feed energy into this high-breaking wave, and the huge mound of water was swept up the entire length of the Chesapeake Bay and the Potomac River.

As the water surged from the bay into the narrowing confines of the Potomac River, a funneling effect resulted in a huge storm surge in Washington (11.3 feet above normal low tide) and Alexandria (twelve feet above normal low tide). This was higher than the storm surge observed at Norfolk (9.3 feet above normal low tide), which was close to the point of landfall. The twelve-foot rise on the Potomac in Old Town, Alexandria, flooded the Torpedo Factory with six feet of water at high tide, while the Washington-Richmond Highway (Route 1) lay submerged under ten feet of water. On the Anacostia River, the devastating tidal surge swept a train off a bridge as it tried to cross, killing ten people. Devastating as it was, there is strong evidence to suggest flooding on the Potomac and neighboring rivers like the Patuxent and Wicomico in Charles County could be even worse.

SLOSH, which stands for the Sea, Lake, and Overland Surges from Hurricanes, is a computer model which estimates a worst-case scenario storm surge based on a number of factors including the strength and speed of a hurricane and its point of landfall. For instance, a strong Category 3 hurricane making landfall in southeastern Virginia, moving NNW at 25 mph, could produce a storm surge as high as 13 feet on the Upper Patuxent River north of Golden Beach, and a storm surge of up to 15 feet on the upper portion of the Wicomico River in southern Charles County. (Bear in mind, these are levels at low tide and do not account for wave action.) However, SLOSH only produces a seven to nine foot storm surge in Washington for such a storm. The 1933 hurricane, in its tropical storm stage, produced a surge more than 2 feet higher. One can only imagine the surge another storm as powerful as Hazel might produce, particularly if it followed the same track as the '33 storm and hit during an astronomical high tide.

Another Spiraling Menace

Severe weather, including short-lived tornadoes, is always a threat with landfalling hurricanes. Tornadoes are spawned as the spiral rain bands that curve outward from the eye wall make their way onshore. As the hurricane moves inland, an intrusion of dry air from the west can wrap into the storm's circulation and lead to more severe weather, especially in the right front

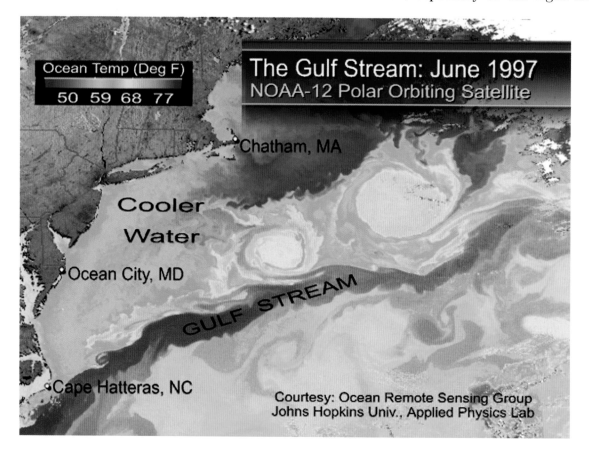

Ocean Temp (Deg F)
50 59 68 77

The Gulf Stream: June 1997
NOAA-12 Polar Orbiting Satellite

Chatham, MA

Cooler Water

Ocean City, MD

GULF STREAM

Cape Hatteras, NC

Courtesy: Ocean Remote Sensing Group
Johns Hopkins Univ., Applied Physics Lab

quadrant of the storm (12 o'clock to 3 o'clock). In 1979, Hurricane David spawned 8 tornadoes in Virginia as it moved up the Atlantic seaboard. One of the twisters hit Fairfax and severely damaged 22 homes. In 1995, a series of 5 tornadoes touched down south and east of D.C. as the remnants of Hurricane Opal passed through the area. One of the twisters, packing winds of 150 mph, caused $5 million in damage in Camp Springs, Maryland.

Freshwater Flooding: The Hurricane's Biggest Threat

Hurricanes have long been feared as coastal killers. In fact, prior to 1970, storm surges that swamped the beaches accounted for 90% of the 25,000 recorded fatalities in the U.S. However, in recent decades, more accurate watches and warnings, coupled with more efficient evacuation of coastal areas have drastically reduced the number of lives lost to the surge. That's the good news. The bad news is that freshwater flooding is putting inland residents' lives more at risk from hurricanes than ever before. Data compiled from 1970 through 2000 indicate that 60% of the U.S. hurricane fatalities were victims who drowned in inland areas due to freshwater flooding. Half of these fatalities resulted when people drowned in their cars or drowned while trying to abandon them. In comparison, hurricane storm surge accounted for only 1% of the hurricane fatalities. These sobering statistics shed a new light on the threat from tropical systems to inland areas, including much of the Washington area.

THE POWER OF WIND:

The destructive power of wind increases dramatically as the wind speed increases. For example, the pressure exerted by wind becomes four times greater as the wind speed doubles. Thus, a 40-mph wind exerts four times the pressure of a 20-mph wind. Consequently, it packs 4 times the destructive force. An 80-mph wind exerts 16 times the pressure of a 20-mph wind!

Data gathered during hurricane reconnaissance flights indicate that wind speeds are often considerably higher just a few hundred feet above the ground, and normally reach their maximum level between 1,000 and 2,000 feet. These stronger winds can be brought down to the surface in quick bursts as the land imparts friction on the wind, creating turbulent wind gusts. Consider Hurricane Bonnie, which rated only a Category 1 storm with 80 to 85-mph surface winds, but had wind speeds measured over 130 mph at an altitude of 300 feet – the height of a 30-story building. Thus, high-rise buildings, particularly near the coast, should not be considered safe havens in hurricanes.

At no time was that threat more evident than in 1972 when Hurricane Agnes tracked from the Florida panhandle up to New York, then turned abruptly south, moving through Pennsylvania into West Virginia. For three days it dumped rain on the Middle Atlantic states. Chantilly, Virginia recorded 16 inches of rain! Agnes triggered widespread flash flooding which caused damage to thousands of homes and businesses, and even the basement of the White House.

Like Agnes, Camille was merely a tropical depression when she produced most of her devastating flooding in 1969. After making landfall in Mississippi as a powerful Category 5 hurricane, Camille weakened as she moved north toward the mountains of southwestern Virginia. As a cold front approached from the west, the moist tropical air was forced to rise over the Blue Ridge Mountains. As band after band of thunderstorms formed, they tracked in succession over the same areas. The deluge resulted in 27" of rain in 8 hours in Nelson County, Virginia, just west of Charlottesville. The resulting flash flooding killed 117 people in Virginia, and remains the state's most deadly natural disaster to date.

Prolific Rainmakers

Both Agnes and Camille were examples of how relatively weak tropical systems can produce incredible amounts of rain. In fact, rainfall amounts of 10-20 inches are common with slow-moving, well-developed systems, particularly along the path of the storm. The reason is that long after the winds die down, the unstable tropical air mass associated with the former hurricane remains ripe for fueling heavy downpours. If any lifting mechanism is encountered, such as an approaching cold front or a natural barrier such as mountains, rainfall may greatly intensify. Rainfall rates of three to four inches per hour, particularly in the mountains, are not unusual. To compound the problem, if the rain follows

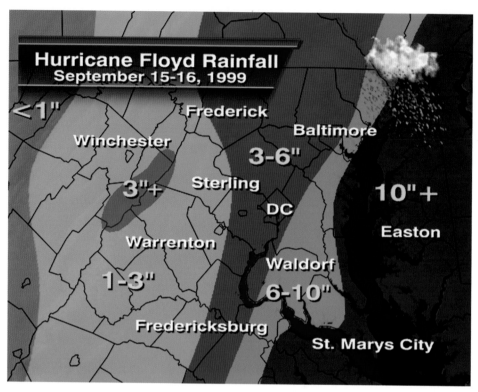

a period of unusually wet weather as Agnes did, the saturated ground behaves much like a wet sponge, and new rainwater cannot be absorbed.

Although much is said about the disastrous consequences of hurricanes, they can produce some positive impacts as well. The next section will explore the beneficial aspects of tropical systems.

Drought Relief

Hurricanes perform a crucial role in nature. On their journey from the tropics to higher latitudes, they transport a tremendous amount of heat and moisture. (It's estimated that the latent heat released by a mature hurricane in one day could supply the United States' electrical needs for about six months.) This energy transport is key to balancing the Earth's heat budget.

On a smaller scale, tropical systems can also bring beneficial rain to areas besieged by drought. The Middle Atlantic Drought of 1998-99 is generally regarded as the second worst of the 20th century. Thousands of acres of crops were lost, res-

ervoirs and wells ran dry, and lawns and shrubs wilted in the intense sun as rainfall deficits approached 20 inches below normal in the Washington area. Mandatory water restrictions were imposed on many area residents, particularly in Maryland, making life rather miserable at times.

Just as the drought conditions became extreme, two September hurricanes finally put an end to the prolonged dry spell. Just after Labor Day, the remnants of Hurricane Dennis brought heavy rain to the Washington area, denting the drought. Then, just three weeks later, Hurricane Floyd made landfall in North Carolina, moved slowly through Eastern North Carolina and continued north over the Delmarva Peninsula. Tragically, the rain, which totaled up to 20 inches in places like Tarboro, North Carolina, and Franklin, Virginia, caused massive flooding. In North Carolina alone, 67,000 homes were flooded. The tropical deluge dumped four to eight inches of rain in the Washington area, with up to a foot of rain falling near Annapolis. The one-two punch delivered by Dennis and Floyd broke the back of the drought, and more importantly, recharged reservoirs and restored near-normal water levels to rivers such as the Potomac and the Rappahannock, which had been running extremely low for several months.

Hurricane Safety

Whether one is battening down the hatches for a hurricane or a blizzard, it's a very good idea to keep a disaster supply kit in the home. The kit should include enough water and nonperishable food for at least three days, flashlights and batteries, a battery-operated radio, a waterproof box for storing important documents, and blankets. Power outages are inevitable in big storms, and so it's wise to fill the car's tank with gas and to pay a visit to the ATM well ahead of the storm. If high winds are expected, anchor loose objects in the yard or bring them inside, consider boarding up windows and glass doors, and be sure to ride out the storm in an interior room with no windows. Of course, the home may not be a safe refuge during a hurricane, particularly if it's a mobile home or is located in a flood-prone area. In either case, evacuate to a certified public shelter or to a location farther inland. Developing a family preparedness plan well in advance of a storm threat will save a lot of last minute confusion and aggravation, and ultimately may be a life-saver.

Visible satellite photo of Hurricane Floyd while it was a category 4 storm off of the coast of Florida, September 13, 1999. Hurricane Floyd made landfall in North Carolina and moved north over the Delmarva Peninsula. Areas of North Carolina received 20 inches of rain, which produced massive flooding. Tropical downpours deluged the Washington area, with up to a foot of rain in Annapolis, Maryland. *NOAA Library*

TROPICAL WEATHER EVENTS

The Powerful Hurricane of September 29, 1896

The Hurricane of September 29, 1896, was one of the most powerful hurricanes ever to impact the Washington area. The storm buffeted the area with hurricane-force winds, causing extensive damage throughout the city and producing twelve fatalities.

The powerful hurricane slammed into the western coastline of Florida during the night of September 28-29, sweeping away the tiny village of Cedar Key, Florida. The storm moved northward, reaching south-central Georgia during the morning of September 29. The storm produced extensive damage in both Gainesville, Florida and Savannah, Georgia, where "nearly every structure was damaged." There were sixty-eight fatalities in Florida, with an additional 25 deaths reported in Georgia.

During the daylight hours of September 29, the tropical storm moved through the Carolinas. By evening, the storm reached southern Virginia, then curved to the left and raced to the north at over 50 mph. In Washington, the southeast wind suddenly jumped from 30 mph to hurricane-force late in the evening of September 29. For the next two hours, the wind was "unparalleled in this part of the country, spreading destruction in every direction." Telegraph wires and city buildings began to succumb to the strong winds. Tree limbs, flying timbers and tin roofs went rocketing through the air. Slate shingles were torn from the tops of houses and carried upon the wind "like birds." Thousands of trees fell – many were snapped off 10-15 feet above the ground. Very few properties escaped having windows blown in or shutters torn off. Many major streets in the downtown area were blocked by fallen debris.

The Washington Weather Bureau forecast the movement of the hurricane very well. Their morning forecast of September 29 called for "threatening weather with brisk to high easterly winds."

For coastal regions, the forecast was for "dangerous gales."

At the official Weather Bureau station, located at 24th and M Streets, the sustained winds reached a peak of 66 mph with gusts to 80 mph. At the Naval Observatory, the anemometer blew away before an accurate peak wind reading could be measured.

The lowest pressure measured at the Weather Bureau was 29.14 inches at 10:50 p.m. on September 29. Rainfall was minimal, with only 0.68 inches measured during the storm. Some lightning and thunder were reported at the height of the storm.

Below are the significant incidents of damage caused by the ferocity of the wind in the Washington area:

- At 1213 Pennsylvania Ave., NW, the west wall of the Albert Building blew out, showering adjoining buildings with tons of brick;
- At 935 Pennsylvania Ave. NW, the Pullman building had its roof blown off and its front wall destroyed;
- At 7th and P Streets, a four-story building had its roof and upper story blown off;
- The steeple of the Presbyterian Church on New York Avenue was toppled and hurled nose first into the ground;
- The tower of the Grand Opera House at the corner of 15th and E Streets was destroyed;
- The roof of the Metropolitan Street Railway Company collapsed, burying 50 streetcars in the debris;
- The Pension Office, Convention Hall and St. Aloysius Church were partially unroofed;
- A wall collapsed at Beatty's Saloon, trapping many inside;
- Rows of homes on H and I Streets had their roofs blown off;
- At local wharves, hundreds of small boats were destroyed.

As the storm raced away from Washington, severe damage was also noted in central Pennsylvania. A bridge over the Susquehanna River blew down and three additional fatalities were noted. Harrisburg recorded a peak wind of 72 mph.

The Chesapeake/Potomac Hurricane of August 23, 1933

The Hurricane of August 23, 1933 is best known for its huge tidal surge up the Chesapeake

---

Speedboat concessions at Hains Point are flooded after the Hurricane of August 23, 1933. The hurricane made landfall near Nags Head, North Carolina and produced 88-mph winds in Norfolk, Virginia. The storm tracked north, just west of Washington. The hurricane pushed a huge tidal surge up the Potomac River that flooded portions of Washington and Alexandria, Virginia under 10 feet of water. *Library of Congress*

Bay and the Potomac River. The hurricane tracked northwest through the Atlantic, passing south of Bermuda on August 21. It made landfall at Nags Head at 4:00 a.m. on August 23, with a central pressure of 28.50 inches. The storm then tracked between Norfolk and Richmond to just west of Washington at 7:00 p.m. on August 23. In Washington, the storm produced 50-mph winds, dropped 6.18 inches of rain, and caused the pressure to fall to 28.94 inches.

The hurricane produced extensive tidal flooding of the Potomac River. A train crossing the Anacostia River was swept off its tracks by the floodwaters, killing ten people. In addition, four people drowned in their cars on the Washington-Baltimore Road when the Little Patuxent River went over its banks. An amusement park in Colonial Beach, located on the Potomac River, was completely swept away. In Alexandria, the Torpedo Factory and the Ford Motor Company were under six feet of water. The

windows were smashed, trees uprooted and roofs were blown off. As Able moved north near Washington, rain pelted down at a rate of over an inch an hour, and winds at National Airport reached a sustained speed of 56 mph, with gusts of 63 mph. Scores of trees were felled in the District and extensive flooding was noted along

Washington-Richmond Highway was submerged under ten feet of water near Alexandria, Virginia, and Bolling Air Force Base was inundated by water up to five feet deep. A total of eighteen fatalities were recorded in the Washington area as a result of the storm.

Hurricane Able Causes Wind and Flood Damage, September 1, 1952

Hurricane Able moved inland near Beaufort, South Carolina, on August 30, 1952, accompanied by winds of up to 100 mph. In Charleston,

Willow trees in East Potomac Park are buffeted by winds during the Hurricane of August 23, 1933. In Washington, winds reached 50 mph and 6.18 inches of rain fell.
Library of Congress

Italian movie star, Gina Lollobrigida, checks out tree damage on the White House lawn after Hurricane Hazel, October 16, 1954. She was in Washington for a courtesy call with President Eisenhower. *Copyright Washington Post; Reprinted by permission of the D.C. Public Library*

Hurricane Hazel – The Windiest Hurricane of the Century, October 15, 1954

Hurricane Hazel holds legendary status in North Carolina, Virginia, and Washington as one of the most severe hurricanes to hit the region. At landfall, Hazel was a strong category 4 hurricane, with a central pressure of 27.70 inches, sustained winds of 135 mph, and wind gusts over 150 mph. Landfall occurred at 9:15 a.m., Friday, October 15, between Myrtle Beach, South Carolina and Wilmington, North Carolina. When Hazel hit the coast, she was speeding north at around 30 mph. Several beach communities in North Carolina were "swept clean" by wind and waves, with virtually nothing left standing.

During the daylight hours of that Friday, Hazel continued to gain forward speed as she

Rock Creek Parkway. In Franconia, a small tornado damaged several homes and wind damage was noted in Potomac, Maryland.

Seneca Creek flooded near Darnestown, Maryland, and became "a lake." A flash flood roared through Ellicott City, Maryland, with ten feet of water, drowning one man and washing away 21 vehicles on Main Street. Parts of Bladensburg, Brentwood, and Riverdale, Maryland were inundated with up to 5 feet of water. Scores of homes were evacuated. To the north, Harrisburg, Pennsylvania suffered extensive damage with winds that were clocked up to 70 mph.

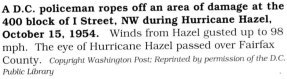

A D.C. policeman ropes off an area of damage at the 400 block of I Street, NW during Hurricane Hazel, October 15, 1954. Winds from Hazel gusted up to 98 mph. The eye of Hurricane Hazel passed over Fairfax County. *Copyright Washington Post; Reprinted by permission of the D.C. Public Library*

continued her damaging rampage through central North Carolina and Virginia. By the time the hurricane reached Fairfax, Virginia, the center had accelerated to an astonishing speed of 60 mph. Hazel's unusually quick movement allowed the storm to arrive in the Washington area while its winds were still above hurricane force. The highest gust recorded at National Airport was 98 mph at 5:05 p.m. on October 15, a record that still stands today (second place is 78 mph).

During the period of peak winds, the control tower at National Airport shook violently and windows blew in. Air control specialists and Weather Bureau employees were forced to abandon their stations fearing for their safety. In addition, several hangars at National Airport were partially unroofed, causing nearby parked cars to be showered with debris.

As the ragged, elongated eye of Hazel passed over Northern Virginia, most areas noted a brief calm. At that time, a pressure of 28.80 inches was recorded at National Airport. As the eye of

Hazel moved away, a terrific squall hit from the northwest with torrential rain and winds of 60 mph. By mid-evening, the skies cleared as the wind rapidly subsided.

In Washington, the rainfall was not particularly heavy. Only 1.73 inches of rain fell during the storm. A drought had been in progress and the rain was considered welcome. During the height of the storm, the rain was quite light with only a warm mist occurring during peak winds. However, the raging southeast winds caused water to back up on the Potomac and spill out of its banks in several locations. Many riverfront buildings were flooded in Alexandria, and Route 1 and

Taping windows at Peoples Drug Store at Thomas Circle in preparation for Hurricane Connie, August 11, 1955. Hurricane Connie made landfall at Morehead City, North Carolina with 100-mph winds. Connie tracked up the Chesapeake Bay, just east of Washington. Winds in Washington gusted to 58 mph. *Copyright Washington Post; Reprinted by permission of the D.C. Public Library*

Hains Point under floodwaters from Tropical Storms Connie and Diane, August 20, 1955. Tropical Storm Diane moved into Washington less than one week after Connie. At National Airport, 10.43 inches of rain fell from both storms. *Copyright Washington Post; Reprinted by permission of the D.C. Public Library*

The Potomac Boat Club at Georgetown is flooded after heavy rain from Tropical Storms Connie and Diane, August 20, 1955. For the month of August, Washington received 14.31 inches of rain.
Copyright Washington Post; Reprinted by permission of the D.C. Public Library

The Anacostia River floods Edmonston, Maryland after heavy rain from Tropical Storms Connie and Diane, August 19, 1955. In Washington, the combined rainfall from both storms was 10.43 inches, which caused widespread flooding.
Copyright Washington Post; Reprinted by permission of the D.C. Public Library

from falling trees and shattering glass. Eight fatalities were reported in Maryland and twelve in Virginia. Over 400,000 local residents lost power. It was estimated that half of the customers in Montgomery and Prince George's Counties lost power with similar disruptions noted in D.C. and Northern Virginia. Three days after the storm,

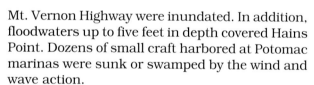

Mt. Vernon Highway were inundated. In addition, floodwaters up to five feet in depth covered Hains Point. Dozens of small craft harbored at Potomac marinas were sunk or swamped by the wind and wave action.

Wind damage in Washington and surrounding areas was extensive. In D.C. alone, at least a half-dozen buildings were partially or totally unroofed by the winds, while others sustained damaged or crumbled walls. Countless trees were ripped apart or felled, blocking streets, crushing houses, smashing cars, and cutting power lines. In the city, nearly every streetcar line was blocked, due to fallen trees and limbs, forcing sanitation employees to work double shifts after the storm to clear the debris. On the Capitol grounds, twenty trees fell, and at the White House, two trees were blown down.

In the immediate Washington area, 39 injuries were reported, with most injuries occurring

Tropical Storm Agnes downed this tree at 30th Street, between K and L Streets in Georgetown, June 22, 1972. Agnes was a minimal category 1 hurricane when it made landfall west of Tallahassee, Florida. Agnes tracked north and produced 49-mph winds in Washington. It was the heavy rain from Agnes, not the winds, that caused most of the damage to the Middle Atlantic region. *Copyright Washington Post; Reprinted by permission of the D.C. Public Library*

there were still over 50,000 people in the area without power.

Hazel continued north and maintained its intensity through the rugged terrain of central Pennsylvania and New York. Hurricane-force gusts were recorded as far north as Ontario. Flash flooding and river flooding were noted from West Virginia to New York. In Canada, flooding was exceptionally devastating. Over a hundred fatalities were reported in the Toronto area.

In all, Hazel did damage in the U.S. totaling at least $250 million and claimed over 90 American fatalities.

Hurricane Connie wrecks the Levin J. Marvel, August 13, 1955

Hurricane Connie, the third storm of the 1955 hurricane season, came ashore at Morehead City, North Carolina, on August 12, 1955. The storm struck with 100-mph winds and 30-foot waves. Compared to Hurricane Hazel, Connie was rather weak. Her central pressure at landfall was a relatively modest 28.40 inches. Nevertheless, the Morehead City area took quite a beating, with extensive damage reported to piers and beach homes.

Unlike Hazel, Connie plodded along at a relatively slow pace of 15

mph. Connie took a more easterly path than Hazel and tracked straight up the Chesapeake Bay. It was not until 6:00 a.m. on August 13 that Connie passed just to the east of Washington, moving between Annapolis and Washington. The lowest pressure recorded at National Airport was 29.11 inches. The wind at National Airport was 49 mph, with gusts to 58 mph. In addition, 6.70 inches of rain fell with Connie.

The major headline with Connie was the sinking of the vacation schooner, the "Levin J. Marvel," in the Chesapeake Bay. The 64-year-old schooner broke apart in the high winds and heavy seas off of North Beach. Thirteen of the twenty-seven on board were rescued after the harrowing experience, but the other fourteen never made it – their bodies were later found washed ashore.

Some trees were blown down with Connie and power outages were commonplace, but compared to Hazel, the damage was relatively minor.

Tropical Storm Diane moved into the area on the heels of Connie, dropping another 1.75 inches of rain from August 16 through August 18. The peak wind associated with Diane was 33 mph. For the month of August, the rainfall total was 14.31 inches, which caused major flooding of the Potomac River.

Piles of logs and trash accumulate in the Potomac River during the flooding of Tropical Storm Agnes, June 24, 1972. Flooding upstream washed debris into the river. The debris is floating next to the docks of the Naval Research Laboratories.
Copyright Washington Post; Reprinted by permission of the D.C. Public Library

Prince William County Supervisor, Vernon Dawson, surveys flood damage to the town of Occoquan, Virginia after flooding from Tropical Storm Agnes, June 1972. These buildings were destroyed by the floodwaters of the Occoquan River. Rainfall at Dulles Airport was 13.65 inches, with some rain totals up to 16 inches in Fairfax County. *Copyright Washington Post; Reprinted by permission of the D.C. Public Library*

A Ford Mustang has been deposited along River Road by floodwaters of Tropical Storm Agnes, June 1972. Sixteen people in the Washington area drowned in the floodwaters of Agnes. *Copyright Washington Post; Reprinted by permission of the D.C. Public Library*

Sandbags at 17th Street, south of Constitution Avenue, are in position to hold back the floodwaters of Tropical Storm Agnes, June 1972. Tremendous flooding of the Potomac River occurred with Agnes, falling just short of the record floods of 1942 and 1936. *Copyright Washington Post; Reprinted by permission of the D.C. Public Library*

Tropical Storm Agnes Creates Historic Flooding, June 21-23, 1972

Tropical Storm Agnes produced monumental rainfall in the Washington area and created some of the worst flooding ever to engulf the region. Agnes began as a tropical depression near the Yucatan Peninsula on June 15. By June 17, Agnes strengthened into a minimal hurricane with 75 mph winds as it trekked slowly northward across the southeast Gulf. Her intensity changed little until landfall was made on the Gulf Coast near Tallahassee, Florida, on June 19. Winds of 60 to 80 mph were reported at landfall.

After landfall, Agnes rapidly weakened. The circulation center moved northeast over land to near Florence, South Carolina, on the morning of June 21. Agnes then moved across eastern North Carolina and tracked north-northeast, toward the Virginia Capes.

In the Washington area, occasional heavy showers began around mid-afternoon of June 21, accompanied by a light northeast wind. During the evening hours, a constant deluge occurred punctuated by nearly continuous lightning and thunder. In a five-hour period nearly five inches of rain fell at National Airport. During the downpour, winds backed to northwest and strengthened to tropical storm force, reaching sustained speeds of 43 mph at National Airport, with gusts as high as 49 mph. Trees and branches fell throughout the area and wires snapped in the gale, cutting power and phones for tens of thousands of homes. However, it was the rain, not the wind that caused nearly all of the death and destruction.

At National Airport, Agnes' 24-hour rainfall total of 7.19 inches nearly broke the all-time record of 7.31 inches set in 1928. Generally, storm totals ranged from 6 inches in the eastern suburbs to as high as 16 inches in Chantilly, Virginia (located in western Fairfax County). Dulles Airport received a storm total of 13.65 inches of rain. In Wheaton, Maryland, the rainfall exceeded one foot.

On the Potomac, the flooding did not exceed the marks set in 1936 and 1942. However, along many of the smaller streams and creeks, the flooding was without precedent. Thousands of homes suffered damage. In Arlandria, Virginia, site of many major past inundations, the flooding was described as the worst ever. The raging Four Mile Run washed out three bridges and closed numerous roads in the area. Seventeen hundred residents were forced to flee their homes as ten feet of water swept through the streets. At the same time, fireman watched helplessly as a fire roared uncontrolled at the Arlandria Shopping Center, destroying 10-12 stores.

In Virginia, portions of nearly every major

artery were closed due to flooding, including Routes 29-211, Route 7, Route 1, Interstate 66 and Interstate 95. Countless secondary roads were likewise affected. A bridge on Route 1 in Woodbridge, Virginia was swept away. At least a dozen other bridges were reported damaged. Flooding was particularly severe in Herndon, Centreville, Manassas Park, Occoquan and Clifton, Virginia. Some towns were virtually isolated.

In Washington, Rock Creek Parkway was closed as abandoned cars were strewn along its length. Likewise, Canal Road and the Whitehurst Freeway were closed, as were parts of Maine Avenue and Independence Avenue.

In Maryland, there were countless flooded roads. A Montgomery Police official was asked for a list of flooded roads. He replied, "you name it, you've got it." Some of the main thoroughfares affected included Georgia Avenue, Sligo Creek Parkway, Jones Bridge Road and Randolph Road. Even the Beltway was closed for several hours in Montgomery and Prince George's Counties, with up to five feet of water reported on the roadway.

As a result of the flooding, police with bullhorns moved through neighborhoods in Aspen Hill, Parkside, Viers Mill Village and Twinbrook on June 22 shouting evacuation orders. Likewise, in Queens Chapel, homes were reported underwater. Extensive flooding was also noted in sections of College Park, Brentwood, Hyattsville and Mount Rainier, Maryland.

Thousands were evacuated in the Laurel, Maryland area during the night as the raging Patuxent spilled out of its banks, ripping out three bridges and damaging several apartments. The business district was engulfed by several feet of water.

The most tragic aspect of this event was the loss of sixteen people in the Washington area who were swept to their deaths in the swirling floodwaters. Most of these drownings involved motorists that were trapped in automobiles.

The winds from Tropical Storm David downed this tree across Rock Creek Parkway, north of Pennsylvania Avenue, September 5, 1979. David hit the Dominican Republic as a category 5 hurricane, with winds up to 172 mph. The storm weakened considerably before brushing the coastline of Florida and making landfall near Savannah, Georgia as a category 1 hurricane. The storm tracked just west of Washington and produced 33-mph winds at National Airport. David also spawned several tornadoes in the Washington area. *Copyright Washington Post; Reprinted by permission of the D.C. Public Library*

As severe as the flooding was in Washington, it was even worse in other places. The James River in Richmond reached an all-time record crest of 36.5 feet, over 27 feet above flood stage. In Pennsylvania, the flooding along the

Susquehanna was unprecedented, breaking previous long-standing high water records by 3-5 feet. Harrisburg, Pennsylvania received 12.55 inches of rain in 24 hours and some areas of Pennsylvania had almost 20 inches of rain. Damages in Pennsylvania alone reached $2 billion, with 250,000 people evacuated from their homes. In all, Agnes claimed 118 lives and did approximately $3 billion in damage.

Tropical Storm David brings Tornadoes and Flooding, September 5, 1979

Hurricane David roared across the Caribbean with catastrophic results during the early part of September 1979. After striking the Antilles, David sliced across the Dominican Republic with winds up to 150 mph. The death toll in the Dominican Republic was almost 1,000. However, the encounter with that rugged island nation also weakened David. The once mighty storm emerged as a Category 1 hurricane. David would never again exceed Category 1 status.

David moved toward Florida, raking the coastline from Palm Beach to Melbourne on September 3. It made a final landfall near Savannah, Georgia, late on September 4. In Florida and Georgia, damage was relatively moderate.

By the morning of September 5, David had been downgraded to a tropical storm as it moved across west central North Carolina. By late afternoon, the center passed just east of Roanoke, and by 9:00 p.m. it moved over Charlottesville, Virginia. Around midnight, David passed near Dulles Airport.

The atmosphere in Washington on September 5 was exceptionally humid. As the storm hit the Washington area, the dew point temperature was 77°F, a tropical value conducive for flooding rain and severe weather. Torrential squalls and strong winds that gusted between 30 and 60 mph encompassed the area. At National Airport, 3.58 inches of rain fell. In many areas, the rainfall totals were far higher; in the immediate eastern and northern suburbs, such as Silver Spring and Capitol Heights, up to 7.0 to 7.5 inches of rain fell.

Tree damage was also noteworthy in parts of the local area. It was most severe in upper Northwest D.C. and in Montgomery County. The National Zoo was closed for a day due to 60 fallen trees. Many major thoroughfares, such as Connecticut Avenue, were littered with fallen trees and downed power lines. Over 140,000 local customers were left without power at the height of the storm.

Several tornadoes touched down across the area. The strongest tornado cut a 25-mile path across the Virginia suburbs. The twister was first sighted at Morningside Drive in Mt. Vernon. The tornado next ripped across the Springfield/Franconia area, where heavy power outages were reported. The tornado then struck the grounds of Woodson High School. The school had been previously hit by the Tornado of 1973. This time, the Woodson stadium took a direct hit. Light poles were "shredded like matchsticks," and the bleachers were damaged. The stadium press box was ripped from its mooring and tossed 500 feet. The residential neighborhood just west of Woodson was also hard hit.

Another small tornado was reported in Anne Arundel County near Crofton, Maryland, where a house was ripped apart; others were reported in Reisterstown (near Baltimore) as well as in St. Mary's and Calvert Counties. In northwest Baltimore, severe flooding badly damaged the Streetcar Museum, a shopping center and a Pepsi bottling plant.

View of floodwaters from the South Ellipse after the Hurricane of August 23, 1933. The hurricane pushed a huge tidal surge up the Potomac River that flooded portions of the Washington area under 10 feet of water. *Library of Congress*

CHAPTER SIX

FLOODS

eandering slowly within its banks, a creek provides one of the most peaceful settings in nature. But given just a few hours, torrential rain can turn a babbling brook into an angry, raging river. The resulting flooding can wash away bridges and roads, sweep away cars, and damage or destroy homes. Floods claim the lives of about 150 people across the country each year – more than hurricanes, tornadoes, and lightning. Flooding occurs in many forms and is caused by a complex set of weather-related and human-related factors.

A Record Deluge

Unionville, Maryland once held a world record for rainfall that stood for nearly 15 years. The record was set on July 4, 1956, when 1.23 inches fell in just one minute! "Cloudbursts" like this are obviously quite rare, but summer thunderstorms routinely produce torrential downpours with rainfall rates of two to three inches per hour in the metro area, and in extreme cases up to five inches per hour, particularly in the mountains. So, what are the scenarios that produce such intense rainfall?

Flooding Takes Many Forms

First, slow moving tropical storms or depressions almost always produce very heavy rainfall, particularly along and just east of the storm's track. Of Virginia's top ten worst river floods in history, seven were attributed to moisture associated with hurricanes, tropical storms, or tropical depressions. Tropical systems originate in waters that are 80°F or warmer, making them extremely rich in both heat and moisture. When this warm, moist air interacts with land, particularly mountains, torrential rain is often the result. One of the Middle Atlantic's worst flooding events also occurred at the hands of a tropical weather system – Agnes, in 1972.

Second, frontal boundaries that stall can set the stage for extremely heavy rain. Occasionally during the summer, fronts will grind to a halt over Virginia and Maryland when winds aloft are light or blowing parallel to the frontal boundary. With nothing to push the front, it simply stops moving. In this situation, warm, moist air riding in from the Atlantic at low levels of the atmosphere will often interact with a southwesterly flow aloft that imports tropical air from the Gulf of Mexico and the Caribbean Sea. As the moisture converges on the front, showers and thunderstorms form, which ring out the atmosphere as if squeezing a wet sponge. In addition, disturbances in the upper atmosphere may also move overhead generating even more rain.

Another common scenario that leads to flooding is called *training*. This is when many intense thunderstorms pass repeatedly over the same area like boxcars behind an engine moving along the railroad tracks. Training occurs as cool downdrafts from thunderstorms rush down to the ground and spread out horizontally. This cool air flow forms miniature cold fronts called *outflow boundaries*. As the warm, moist air is pushed up by the outflow boundary, more thunderstorms are generated in the same area. These new cells create more downdrafts that reinforce the outflow boundary, and the process is repeated.

Heavy Rain is Only Half of the Flooding Equation

It's not only how hard it rains, but also the duration of the rainfall that determines the likelihood of flooding. For example, a thunderstorm that is dumping rain at the rate of two inches per hour but zipping along at 40 mph may create some large puddles on roadways, but major flooding would be very unlikely. However, if the same thunderstorm were to remain stationary over an area for two hours, the resulting four inches of rain would create an entirely different situation.

Another factor affecting the likelihood of flooding is topography. First, mountains act as a physical barrier to transient air masses. They often enhance rainfall as the air is forced to rise over them. The steep terrain also accelerates rainfall runoff into the valley below. Sometimes, narrow rivers and streams cannot handle the enormous volume of water, and consequently overrun their banks. Second, the extent of flooding is highly dependent on the ground conditions. Highly developed urban areas like Washington resemble "asphalt jungles," with thousands of miles of roads, driveways, and countless parking lots forming an impervious barrier to rainfall. Following a heavy rain, runoff into rivers, creeks, and streams from these paved surfaces are up to six times over what would occur on natural terrain. The result is termed *urban flooding*. City streets become like streams, and underpasses can flood in minutes stranding people in their cars. Flooding becomes even more likely when storm drains or culverts are blocked by debris. The District of Columbia and the nearby suburbs are quite prone to urban flooding.

Soil conditions also play a large role in flooding by affecting how much rain is soaked up by the natural landscape. Sandy soil will absorb moisture more readily than clay soil – that is why it usually takes more rain to trigger flooding on the sandy soil of the Eastern Shore than it does on the compact clay soil found in many areas west of Washington. Also, if the ground is saturated as a result of days or weeks of rain, or frozen due to several nights of subfreezing temperatures, much of the rain will run off rather than seep into the ground. In this case, it may only take an inch of rain falling in less than an hour to trigger flooding.

A Full Moon Can Spell Trouble

The Potomac River starts in the mountains of western Maryland and flows nearly 200 miles to the southeast where it empties into the Chesapeake Bay. The Potomac is unique in that it is subject to two different kinds of flooding – *tidal flooding* and *river flooding*. The tidal portion of the Potomac, south of Chain Bridge, is subject to tidal flooding. Tides are produced by the gravitational forces that the moon, and to a lesser extent the sun, exert on the surface of the Earth. Tidal flooding is caused by higher-than-normal tides generated by tropical storms, hurricanes or slow-moving nor'easters. These storms often create strong, persistent winds blowing out of the east that "bottle up" the lower Chesapeake Bay causing water to "backwash" up the Potomac. This can produce extremely high tides.

Flooding is compounded if the storms hit during astronomical high tides called *Spring Tides*. In this case, "spring" has nothing to do with the season, but refers to the fact that the tides "spring up" during certain times of the month. Spring tides occur during the new and full phases of the moon, when the moon and the sun are nearly lined up with the Earth. During these phases, the gravitational pulls exerted by the moon and the sun on Earth are acting in the same direction, which produces higher water levels at high tide, and lower water levels at low tide.

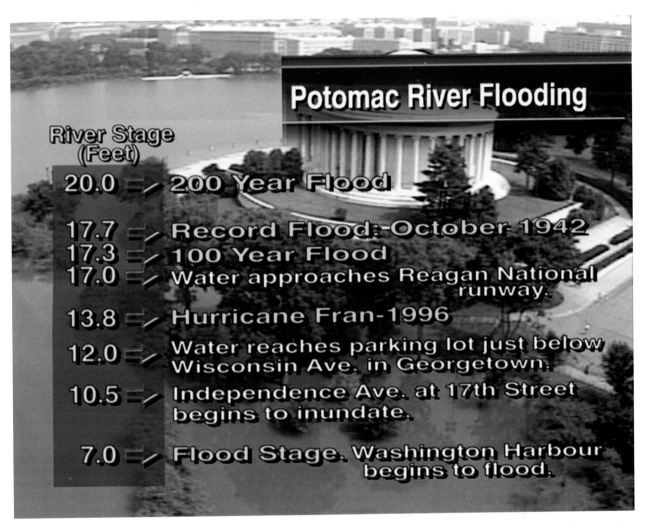

Potomac River Flooding

River Stage (Feet)	
20.0	200 Year Flood
17.7	Record Flood - October 1942
17.3	100 Year Flood
17.0	Water approaches Reagan National runway.
13.8	Hurricane Fran - 1996
12.0	Water reaches parking lot just below Wisconsin Ave. in Georgetown.
10.5	Independence Ave. at 17th Street begins to inundate.
7.0	Flood Stage. Washington Harbour begins to flood.

River Flooding

The Potomac River, as well as the Rappahannock and Shenandoah Rivers in Virginia, are all susceptible to *river flooding*. *River flooding* is usually the result of days of heavy rain falling over a large portion of the river basin and watershed. In some cases, runoff from heavy rain may flow hundreds of miles downstream before it causes flooding. River floods can result from extremely heavy downpours associated with decaying tropical systems, from heavy rain in winter or spring falling on snow-covered ground, or from massive piles of ice called *ice jams* that cause water to back up. Just two weeks after the Blizzard of 1996 dumped two to four feet of snow on the Washington area, 60-degree temperatures and

heavy rain led to the "Big Melt." Flooding on the Potomac River damaged homes and businesses from West Virginia to Alexandria, and wiped out 80% of the paths and bridges in the C&O National Historic Park.

Water, Water Everywhere

One of the worst river floods in the region's history occurred back in October of 1942. During a four-day deluge from the 12th to the 16th, over a foot of rain fell on the upper Potomac near Paw Paw, West Virginia, while up to 19 inches of rain poured down from the skies along the Skyline Drive in western Virginia. Fredericksburg was ravaged by flooding when the Rappahannock

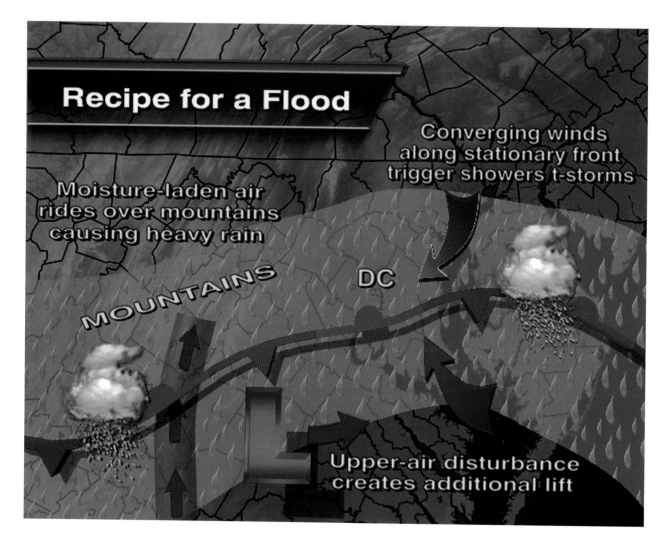

Recipe for a Flood

Moisture-laden air rides over mountains causing heavy rain

Converging winds along stationary front trigger showers t-storms

MOUNTAINS

DC

Upper-air disturbance creates additional lift

River crested at 42.6 feet – 24 feet above flood stage. (Flood stage is the height of the river when property damage begins.) Front Royal, on the South Fork of the Shenandoah, also suffered major damage when that river crested nearly 23 feet above flood stage.

As the wall of water flowed downstream on the Potomac, a flood crest of 17.7 feet was measured in Washington – 10.7 feet above flood stage. (The "flood crest" is the highest stage the floodwaters reach, and moves like the crest of a wave down the river.) To put the Potomac's flood crest of 17.7 feet in perspective, Washington Harbour begins to flood at a level of seven feet; at 13.5 feet, water reaches Wisconsin Avenue and K Streets in NW Washington; and at 17 feet, water

approaches the runway at Ronald Reagan Washington National Airport.

A level of 17.3 feet on the Potomac is considered a 100-year flood for Washington. This means there is a one percent chance of such a flood occurring in any given year. Statistics, though, can be a bit misleading – Washington has endured two "100-year floods" since 1936.

The Power of Water

Fortunately, water levels on the Potomac rise too slowly east of the mountains to trigger what is called *flash flooding*. However, many of the Potomac's tributaries, as well as smaller creeks and streams in the Washington area, are periodically subject to flash flooding. Just as the name implies, flash floods occur in a "flash," with little or no warning. Under normal circumstances, 2 to 2.5 inches of rain in one hour, or three to four inches in a three-hour period will trigger flash flooding in many low-lying areas east of the Blue Ridge. These amounts may be much lower, or even slightly higher, depending on a number of factors including the degree to which the soil is saturated, the slope of the terrain, and the vegetation covering the ground.

Flash floods are especially deadly because they flow at high speeds and produce rapid rises in water levels, often catching people unprepared. They become even more dangerous at night, particularly when roads are flooded or washed out by the torrents. Eighty percent of flood deaths in the United States involve vehicles, and most are motorists trying to navigate through floodwaters. This is an extremely risky proposition considering the amazing power of water: for each foot the water rises up the side of the car, a force of 1,500 pounds is exerted upward on the bottom of the car. Given the weight of the average car, this means *two feet of water will carry away most automobiles.* In fact, six inches of moving water will sweep most people off their feet.

In many western states, it takes as little as a half inch of rain to trigger flash flooding. Las Vegas, Nevada, which averages only 4 inches of rain annually, saw over 2 inches in a 90-minute period on November 9, 1999. Local authorities performed 200 swift water rescues during the flash flooding that ensued. Miraculously, the floods only claimed two lives but did cause $20 million dollars in damage.

Cars are swept into a ditch after a parking lot collapses in Glen Echo Park, May 5, 1989. Flash flooding from 5 inches of rain claimed three lives in the Washington area that day. *WJLA*

Washington's Flood Prone Areas

The relatively flat terrain, better drainage systems, and relatively porous soil make the suburban Washington region less vulnerable to flash flooding than most areas in the western U.S. However, there are a few notorious trouble spots that have been the sites of numerous rescues and drownings over the years. They include Rock Creek in the District, the Northwest Branch of the Anacostia River in Prince George's County; Seneca Creek in Montgomery County, the Monocacy River in Frederick County, Four Mile Run and Difficult Run in Fairfax County, and Goose Creek in Loudoun County. Many of these sites lived up to their reputation on May 5, 1989 when up to five inches of rain fell in Washington and its suburbs. The raging floodwaters killed three people, inundated hundreds of homes and businesses, and washed out a parking lot in Glen Echo Park, sweeping 63 cars into a ditch.

The potential for devastating floods looms much larger in the mountains of western Virginia, western Maryland, and West Virginia. One of the worst flash floods in Virginia history occurred on June 27, 1995 when a front stalled near the mountains. Up to 20 inches of rain fell on the eastern slopes of the Blue Ridge in Madison County, Virginia, about 80 miles southwest of

Washington. Torrential rain produced a 500-year flood crest on the Rapidan River. (A 500-year flood means there is a 1 percent chance of it occurring in any five-year period.) The massive deluge triggered landslides, washed away roads, and destroyed entire farms. Eighty people had to be rescued – many from their rooftops. Miraculously, only three people were killed in Virginia.

Flood Safety

Staying safe during times of flooding can be summed up in two statements: (1) never underestimate the power of water, and (2) when floodwaters threaten, head for higher ground. If you are in your home and you are advised to evacuate, do so immediately. Fortunately, most Washington area residents do not live in flood-prone areas. However, drivers and passengers in cars face potentially life-threatening situations if attempting to drive through high water. Remember, driving over flooded roadways is extremely dangerous – not only is it often impossible to judge how deep and how fast the water is moving, but the roadbed itself may have been washed away.

Because children are curious by nature, they

Flash flooding triggered a landslide that nearly buried this home in Madison County, Virginia, June 27, 1995. Up to 20 inches of rain fell in Madison County, causing a 500-year flood crest on the Rapidan River. *WJLA*

should be watched closely when flooding is in progress. Rivers, creeks, streams, drainage ditches, and storm drains are all prone to rapid water rises and should be off limits to children.

Finally, plan ahead. Know your area's flood risk and identify at least two possible locations you could go if ordered to evacuate. Purchasing flood insurance may also be a very wise investment considering most homeowner's and renter's insurance policies do not cover flooding.

FLOOD EVENTS

The Great Flood of 1889

A major storm system moved north from the Gulf of Mexico, triggering a massive rainstorm along the East Coast on May 30-31, 1889. In Washington, the rain total was 4.4 inches. The same storm system was responsible for the Johnstown Flood that claimed over 2,000 lives in Pennsylvania.

The center of the storm passed through western Virginia and produced heavy thunderstorms throughout the region. The rainstorm culminated a very moist month, with the monthly rainfall in Washington totaling 11.78 inches.

The Potomac River reached its crest in Washington on June 2, 1889, at 11.5 feet above flood stage. This holds the unofficial flood record for Washington. Only the floods of 1936 and 1942 have approached this level. (The flood of 1889 is not included in modern flood records. The Flood of 1942 holds the official Washington flood record at 10.7 feet above flood stage.)

The swirling flood invaded the city's downtown section to a depth of several feet. The entire length of Pennsylvania Avenue was covered with one to four feet of water. The Long Bridge crossing the Potomac was battered relentlessly throughout this episode. It survived the worst of the flood, but then partially collapsed as the waters receded.

Violent and deadly weather accompanied this storm. In Danville, Virginia, a six-story brick building collapsed in high winds, resulting in five deaths. In Falling Waters, West Virginia (about ten miles southwest of Hagerstown, Maryland), a powerful tornado struck that claimed two lives.

This storm will forever be remembered for triggering the catastrophic Johnstown, Pennsylvania Flood, one of the greatest

An illustration showing floodwaters from the Flood of June 2, 1889. On May 30-31, a moist storm from the Gulf of Mexico dropped 4.40 inches of rain in Washington. The total rainfall for May in Washington was 11.78 inches. *Library of Congress*

Pennsylvania Avenue is under floodwaters from the Potomac River during the Flood of June 2, 1889. The entire length of Pennsylvania Avenue was covered with one to four feet of water. The catastrophic Johnstown Flood in Pennsylvania occurred two days earlier. *Library of Congress*

Pennsylvania Avenue during the Flood of June 2, 1889. The Potomac River crested 12.5 feet above flood stage, flooding many areas of Washington. *Library of Congress*

A scene in Washington during the Flood of June 2, 1889. Boats shared the roads with horse-drawn carriages after a storm dropped 4.40 inches of rain on Washington. *Library of Congress*

An illustration of the flooded railroad depot on Sixth Street and Pennsylvania Avenue during the Flood of June 2, 1889. The Flood of 1889 was of the same magnitude as Washington's record Flood of 1942. *Library of Congress*

disasters in American history. Approximately 8 inches of rain fell during the overnight hours of May 30-31, which triggered the sudden collapse of the South Fork Dam on May 31, 1889. This dam was thought to be the world's largest earth dam – 72 feet high, 300 yards wide. The ensuing wall of water and debris claimed between 2,000 and 3,000 lives.

The Flood of 1924

Heavy rain fell along the Potomac watershed on May 7-8, 1924 and again on May 11, 1924. The runoff triggered the worst Potomac River flood

Chain Bridge, over the Potomac River, is barely above floodwaters, May 13, 1924. The bridge survived this flood but later succumbed to the Flood of 1936, which produced floodwaters that were six feet higher. *Library of Congress*

since May 1889. In Washington, the crest was 4.2 feet above flood stage. Damage was relatively minor in Washington, but the C&O Canal suffered extensive damage. After the flood, maintenance and repair of the canal would cease. Never again would the C&O be a viable business.

The Great Potomac Flood of 1936

On March 17, 1936, an intense storm system moved north from the southern states. The Potomac River was already swollen from recent rain and snowmelt. This disturbance produced additional heavy rain, particularly around the headwaters of the Potomac River, where rain totals reached 5 inches in West Virginia.

The Potomac River's flood crest in Washington was over ten feet above flood stage. The floodwaters lapped at the national

A washout of the C&O canal into the Potomac River above Georgetown, May 13, 1924. Railroads had started to take business away from the canal prior to 1924. Extensive damage to the canal by the Flood of 1924 ensured that the C&O would never again be a viable business. The Federal government acquired the canal in 1938, and the Chesapeake and Ohio National Historic Park was formed in 1971. *Library of Congress*

Floodwaters at Great Falls, May 13, 1924. After the Flood of 1924, maintenance and repair of the C&O canal ceased. Today, the 184-mile towpath is maintained for hiking, running and bicycling as part of the Chesapeake and Ohio National Historic Park. *Library of Congress*

monuments and put the runways of National Airport underwater. In Bladensburg, flooding of the Anacostia River reached first floor windows. Fortunately, Washington's downtown area was spared, largely due to the heroic sandbagging efforts of 2,000 Civil Conservation Corps workers, who constructed a four-foot dike. The Civil Conservation Corps was one of President Roosevelt's depression relief efforts to hire the

The Record Flood of 1942

Strong high pressure to the north of D.C. and a weak storm to the south put Washington in persistent, moist northeast flow during mid-October, 1942. From October 13 to October 17, 6.27 inches of rain doused the city. Even heavier amounts fell in Virginia's mountains, with 12.3 inches of rain pelting Big Meadows in a 24-hour

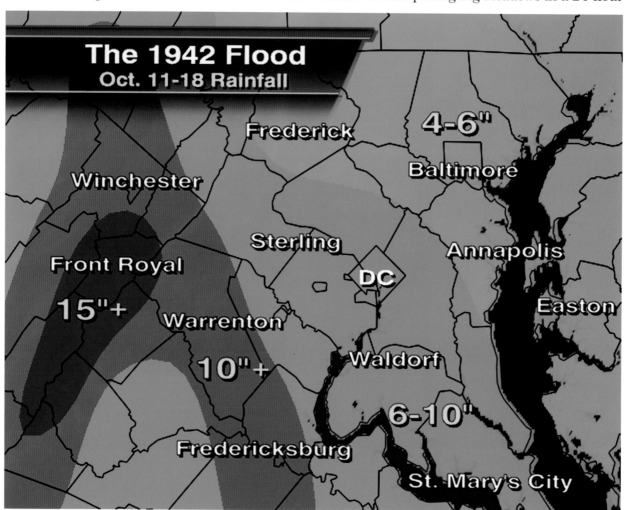

unemployed to work on government-funded projects.

The 1936 flood was extremely devastating in Pennsylvania and parts of New England. Over 150 deaths resulted. Many flood crest records were established – some of which still stand today.

period. The resulting crest on the Potomac River was 17.7 feet in Washington, which was 10.7 feet above flood stage.

The flood of 1942 was one of the region's largest floods and holds the official record for D.C. Similar to the 1936 flood event, there was an

Floodwaters at 31ˢᵗ and K Streets, NW, October 17, 1942. The Flood of 1942 is considered Washington's greatest modern flood. The Potomac River crested to 17.7 feet, which is 10.7 feet above flood stage. *Washingtoniana Division, D.C. Public Library*

Floodwaters reach to the steps of the Jefferson Memorial, October 17, 1942. Washington's rainfall was 6.27 inches, but 10 to 15 inches of rain fell to the west of D.C. *NOAA Library*

Flooded monument grounds looking up 17ᵗʰ Street, October 17, 1942. Sandbagging efforts were necessary in many parts of Washington to hold back the floodwaters. *Washingtoniana Division, D.C. Public Library*

U.S. soldiers and D.C. firemen evacuate Washington residents from their flooded homes, October 17, 1942. Rescue efforts occurred in low-lying areas of the city. *Library of Congress*

A duck seeks refuge on a stool during the flooding from Hurricane Hazel, October 16, 1954. Tidal flooding of the Potomac River, caused by Hazel's strong southeast winds, flooded this store.
Copyright Washington Post; Reprinted by permission of the D.C. Public Library

extensive sandbagging effort to hold back the floodwaters in low-lying areas of Washington.

The Flash Flood of July 22, 1969

A cluster of slow-moving thunderstorms passed through the area during the evening of July 22, 1969, producing 4.38 inches of rain at National Airport. Of that amount, 3.29 inches fell in one hour. In Vienna, 7.52 inches of rain fell. The storms struck hard in Fairfax, Arlington, Prince George's County, and the District, but left Montgomery County relatively untouched.

The flooding was particularly devastating in Arlandria (between Arlington and Alexandria) where the water level in Four Mile Run reached an estimated 19 feet during the storm. This level broke the previous record of 17 feet set in another flood which took place in August 1963. The flooding was so severe that the 10-foot water gauges which measure flood levels around Route 1 and Mount Vernon Avenue were completely submerged.

The Arlandria Shopping Center suffered major damage with many store windows having been broken by the rushing flood waters. The

The underpass below the Anacostia Freeway fills with water, September 12, 1960. The flooding resulted from the rain associated with Hurricane Donna. The storm dropped 2.87 inches of rain on Washington. *Copyright Washington Post; Reprinted by permission of the D.C. Public Library*

New cars were washed down Four Mile Run from the Rosenthal Auto Dealer and deposited along the stream's bank after the Flash Flood of July 22, 1969. Four Mile Run had an estimated water level of 19 feet above bank-full during the flood. Five years later, Congress authorized $40 million to build a flood control channel to the Potomac River which has helped alleviate Four Mile Run's flood problems. *Copyright Washington Post; Reprinted by permission of the D.C. Public Library*

A station wagon and utility pole fell into a sinkhole that was created when Sligo Creek flooded the parkway, July 22, 1969. A slow-moving cluster of thunderstorms dumped up to a foot of rain in a very short period of time. *Copyright Washington Post; Reprinted by permission of the D.C. Public Library*

The Bailey's Crossroads Fire Department performs search and rescue missions, July 22, 1969. The ten-foot flood gauge at Route 1 and Mount Vernon Avenue was completely submerged after seven inches of rain fell in Fairfax County. *Copyright Washington Post; Reprinted by permission of the D.C. Public Library*

Floodwaters rush through the historic town of Occoquan, Virginia after Tropical Storm Agnes, June 24, 1972. The town suffered major damage from the flooding of the Occoquan River. *Copyright Washington Post; Reprinted by permission of the D.C. Public Library*

Floodwater blasts a pipeline span over the Occoquan River after Tropical Storm Agnes, June 22, 1972 (above) and the same pipeline span during low water levels (right). The pipeline span is located just upstream from the town of Occoquan, Virginia. Farther downstream, the flooding washed out a section of the Route 1 bridge over the Occoquan River. Rainfall at Dulles Airport was 13.65 inches, with 16 inches of rain measured in Chantilly, Virginia.

Copyright Washington Post; Reprinted by permission of the D.C. Public Library.

Ellicott City, Maryland is under water after Tropical Storm Agnes, June 1972. Ellicott City has a long history of flooding from the Patapsco River, but the Agnes Flood was one of its worst. The historic Ellicott House was destroyed and the nation's oldest railroad station was damaged. After the flood, a citizens group restored the railroad station into a museum. *Ruby Tuesday, Columbia.*

Torrential rain partially buried the front end of this car in mud and leaves, December 2, 1974. Rainfall at National Airport was 1.90 inches, with 1.57 inches falling in two hours. *Copyright Washington Post; Reprinted by permission of the D.C. Public Library*

flood waters also swept the car lots of Cherner Motors and Rosenthal Chevrolet in Shirlington, piling cars into one another and sweeping some into Four Mile Run. In all, twenty cars were found in Four Mile Run after the flooding subsided.

The storm struck just before the scheduled start of the Major League All-Star baseball game at RFK Stadium. More than 40,000 fans, including President Richard Nixon, were caught in the sudden cloudburst. The game was postponed as the dugouts filled chest-deep with water.

As the storm cluster moved southeast away from the District, it produced even heavier rain in southern Maryland, with Leonardtown, Maryland, recording an amazing 12.44 inches.

The Agnes Flood of June 21-23, 1972

Tropical Storm Agnes produced monumental rainfall in the Washington area and created some of the worst flooding ever to engulf the region. At National Airport, Agnes' 24-hour rainfall total of 7.19 inches approached the all-time record of 7.31 inches set in 1928. Generally, storm totals ranged from 6 inches in the eastern suburbs to as high as 16 inches in Chantilly, Virginia (located in western Fairfax County). Dulles Airport received a storm total of 13.65 inches of rain. In Wheaton, Maryland, the rainfall exceeded one foot.

On the Potomac, the flood level was 15.5 feet in D.C., which was over eight feet above flood stage. The Agnes Flood ranks fifth in Washington for the greatest river flooding event. However, along many of the smaller streams and creeks, the flooding was without precedent. Thousands of homes suffered damage. In Virginia, portions of nearly every major artery were closed due to flooding. In Maryland, there were also countless flooded roads.

Some of the worst flooding occurred in Richmond, Virginia, when the James River reached an all-time record crest of 36.5 feet, over 27 feet above flood stage. In Maryland, flooding from the Patapsco River caused significant damage to historic buildings in Ellicott City. In Pennsylvania, the flooding along the Susquehanna was unprecedented, breaking previous long-standing flood records by 3 to 5 feet. Harrisburg received 12.55 inches of rain in 24 hours, with some rainfall totals approaching 20 inches. Damages in Pennsylvania alone reached $2 billion and 250,000 people were evacuated from their homes. In all, Agnes claimed 118 lives and did approximately $3 billion in damage. (Note: See chapter 5 for more on Agnes.)

POTOMAC RIVER FLOODS

On the average, the Potomac River floods approximately once every two years, most frequently during the spring and least frequently during the summer. Heavy rain falling in Virginia, Maryland, and West Virginia have historically caused the greatest floods in the area. Washington's record flood occurred on October 17, 1942, when heavy rain swelled the Potomac River to a level of 17.7 feet – more than 10 feet above flood stage.

Five days of steady rain caused another major flood of the Potomac River on June 2, 1889. The floodwaters covered the entire length of Pennsylvania Avenue with water depths of one to four feet. In Washington, the Flood of 1889 reached 19.5 feet above flood stage, setting the city's unofficial flood record. (In 1889, flood readings were made at the Aqueduct Bridge in Georgetown. The location for flood readings has moved several times since 1889, and the gauge is presently located at the base of Wisconsin Avenue in Georgetown.)

Flooding is not always the result of heavy rain. Ice jams on the Potomac River can lead to flooding in Washington. During February 1881, a huge ice gorge caused a flood, which produced 40 inches of water on Constitution Avenue at 15th Street. Large ice jams also occurred on the Potomac River in February 1918 and February 1948.

Another form of Potomac River flooding is tidal flooding. Strong, persistent winds can push water up the river, dramatically increasing the level of the tides. An example of tidal flooding occurred during the Hurricane of 1933. Strong southeast winds produced a high tide of eleven feet in Washington (normal high tides are about three feet), flooding many areas of the city. Ironically, the rain associated with the hurricane produced very little flooding upstream from Washington, in the stretches of the river not affected by tides.

Rapid snowmelt can also lead to flooding of the Potomac River. During January 1996, a two-foot snowpack melted almost overnight during a rainstorm, leading to a rapid rise in the Potomac River. Extreme flooding occurred from West Virginia to Washington. Similar flooding occurred in March 1936, when snowmelt and heavy rain caused a near-record flood in Washington.

The two tables below show flood levels of the Potomac River. The first table lists flood levels at Little Falls, located ten miles downstream from Great Falls and just upstream from Chain Bridge. The second table lists flood levels at Washington, with the flood gauge located in Georgetown.

Floods at Little Falls (flood stage = 10 feet)		Floods at Washington (flood stage = 7 feet)	
March 19, 1936	28.1 feet	June 2, 1889	19.5 feet*
October 17, 1942	26.9 feet	October 17, 1942	17.7 feet**
April 28, 1937	23.3 feet	March 19, 1936	17.3 feet
June 24, 1972	22.0 feet	June 24, 1972	15.5 feet
January 21, 1996	19.3 feet	April 28, 1937	14.3 feet
November 7, 1985	18.0 feet	January 21, 1996	13.9 feet
September 8, 1996	17.8 feet	February 14, 1918	13.8 feet
August 20, 1955	17.6 feet	September 8, 1996	13.7 feet

*Unofficial Record
**Official Record

The Flood of 1985

The flood of 1985 resulted from extremely heavy rain along the Appalachian Mountains as two storm systems hit the area in quick succession. First, the remnants of Hurricane Juan moved into the area on November 1, dropping significant rain. Soon after, a storm system developed in the Gulf of Mexico and moved north into Virginia on November 4 and 5. Rainfall became extremely heavy along the eastern Appalachian Mountains, particularly in Central and Northern Virginia. Devastating flash floods occurred in the mountain regions, which later led to terrible river flooding in Central and Northern Virginia, including severe flooding of the Potomac, Rappahannock and James Rivers.

Tidal flooding was also a factor with both storm systems. A prolonged wind from the southeast led to tidal flooding of the Chesapeake Bay and tributaries. However, the tidal flooding was not nearly as severe as the flooding that occurred upstream from the heavy rain.

The Floods of 1996

Two major flooding events of the Potomac River occurred in 1996. The first and largest of the two flood events occurred in January, and the second and smaller flood event occurred in September.

The Flood of January 1996 occurred after a heavy rainfall of two to five inches combined with a quick snowmelt. The resulting runoff flooded the Potomac River and many low-lying areas in

The Potomac River floods an entrance road in Great Falls Park, January 21, 1996. Floodwaters came within 9 feet of the Great Falls Visitors Center. The Flood of January 1996 occurred from a heavy rainfall combined with a rapid snowmelt. *National Park Service*

The Flood of January 21, 1996 at Great Falls (above) and the normal water level at Great Falls (right).
The Potomac River rose 85 feet in 48 hours at Great Falls during the Flood of January 1996. The flood marker shows how the Potomac River floods have compared at Great Falls, which is located about 15 miles upstream from Washington. The Flood of January 1996 is the fifth largest flood in the past 100 years.

National Park Service

Flood waters at an overlook at Great Falls Park, Virginia, September 8, 1996. Floodwaters have risen to the top of a 75-foot overlook on the Potomac River at Great Falls. On September 6, 1996, remnants of Hurricane Fran produced very heavy rain to the west of the Washington area. Frostburg, Maryland received 6.53 inches of rain while only 1.55 inches fell at National Airport.

National Park Service

Potomac River floodwaters at Great Falls Park, September 8, 1996. This fence is usually high above the Potomac River and helps hikers avoid dangerous terrain.

National Park Service

Virginia, Maryland and Washington. The flooding caused extensive damage along the C & O canal towpath. The quickly rising waters also trapped one individual on the rocks near Great Falls who had to be rescued by helicopter.

The setup for the Flood of January 1996 occurred when heavy snows fell in the Middle Atlantic during the early part of the month. From January 6 to January 14, approximately two feet of snow fell in a wide area across Northern Virginia, Central Maryland and Northern West Virginia. The water content of the snow pack was approximately two to three inches. On January 18 and 19, a rapid warmup occurred that was accompanied by a heavy rainfall of two to five inches. Almost the entire snowpack melted within a 24-hour period as the rain fell. The previously frozen ground did not have much time to thaw and absorb the moisture. A combined total of four to eight inches of water quickly drained into streams and rivers, creating a very rapid flood. The following statistics measured near Little Falls show how quickly the Potomac River rose during the flood:

Jan 19 - 24,900 cubic feet/second of water
Jan 20 - 178,300 cubic feet/second of water
Jan 21- 347,200 cubic feet/second of water
Jan 22- 143,100 cubic feet/second of water

In D.C., the Flood of January 1996 crested 6 to 7 feet above flood stage. At Great Falls, the Potomac River rose 85 feet in 48 hours. The Flood of January 1996 is Washington's fifth largest flood in the past 100 years, ranking behind the floods of 1936, 1937, 1942 and 1972.

Ironically, another major flood occurred during the same year. On September 6, 1996, remnants of Hurricane Fran produced very heavy rain

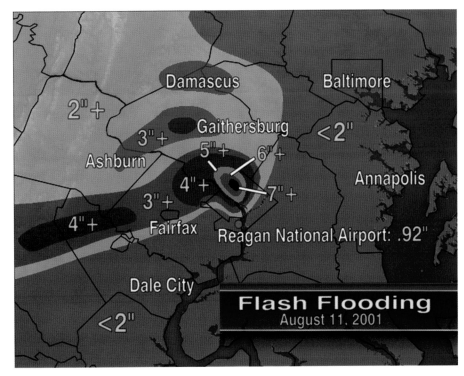

to the west of Washington. Frostburg, Maryland, received 6.53 inches of rain, Dulles Airport received 3.62 inches of rain, and National Airport received 1.55 inches of rain. The ensuing flood was not quite as severe as the January flood, but still caused severe damage along the Potomac River. At Little Falls, the Flood of September 1996 equaled the severe Flood of 1985.

The Extreme Flash Flood of August 11, 2001

The extreme flash flooding of August 11 occurred along a narrow band from Warrenton, Virginia, through Northwest Washington. The flooding was some of the worst in memory for many areas of D.C. and surrounding suburbs. Many streets were heavily flooded and the force of overflowing storm sewers pushed up manhole covers in boiling torrents of water. Countless basements filled with water and flood-related power outages lasted for days. Some streets remained closed for almost a week. After the flood, Washington was declared a Federal disaster area.

The flooding was related to slow moving

thunderstorms that dropped very heavy rain. On August 10, the first bout of thunderstorms struck the area. Over two inches of rain fell in Northwest D.C., which resulted in minor street flooding. A day later, on August 11, a broad area of thunderstorms formed to the west of Washington and moved very slowly through the area. Over seven inches of rain fell in NW Washington. The combined two-day rain totals were over nine inches in parts of D.C. In Fairfax County and Montgomery County, 5-inch rainfall totals were common. Meanwhile, SE Washington totaled less than two inches of rain with minimal flooding.

The flood damage in Washington was described as the worst in more than fifty years. The hardest hit areas of the District were low spots near Rhode Island Avenue and First Street, NW. Sewer lines could not handle the quick runoff from the slow moving thunderstorms and thousands of gallons of raw sewage gushed up from under city streets, covering roads and seeping into homes and businesses. Thousands of properties suffered flood damage. Many basements were filled with water, sewage, and, in some instances, dead rats from the flooded sewer system.

Street flooding in Washington during the Flood of August 11, 2001. The flash flooding of August 11 occurred along a narrow band from Warrenton, Virginia through Fairfax County, and extended into northern D.C. Up to 7 inches of rain fell on parts of Washington. *WJLA*

A large sinkhole near First Street, NW, August 12, 2001. Sewers could not handle the quick runoff from the slow moving thunderstorms of August 11. This sinkhole resulted when the sewer line washed out. *Photo by Dudley M. Brooks ©2001,*
The Washington Post, Reprinted with permission

Frying eggs on the Capitol steps, August 31, 1947. These young ladies have come equipped with a plate, spatula, and salt shaker. Washington's high temperature reached 94°F that day. *Copyright Washington Post; Reprinted by permission of the D.C. Public Library*

CHAPTER SEVEN

HEAT WAVES

ummer in the nation's capital can be unbearably hot and humid. The summer of 1999 will be remembered not only for the searing heat but also for the terrible drought, which dried up lakes and reservoirs, and baked lawns and gardens to a crisp. The heat peaked in July as the mercury soared to at least 90°F on 22 days, while only 1.01 inches of rain fell at Reagan National Airport. In fact, July 1999 went in the record book as the second warmest month of all time in Washington, and the third-driest July since 1871.

Heat is a Killer

In most years in the U.S., severe heat claims more lives than hurricanes and tornadoes combined. A nationwide average of about 200 people die each year due to the direct results of blistering heat and oppressive humidity. The majority of victims are either very young or very old – their bodies can't cope with the prolonged stress brought on during these *heat waves.*

The term *heat wave* refers to an extended period of very high temperatures and humidity, ranging from a few days to perhaps a week or longer. The 24-day stretch from July 18 through August 10, 1930, remains the longest and most intense heat wave on record in the Washington area. During that period, the District tied its all-time record high of 106°F and hit the century

mark an amazing eleven times! In Frederick, Maryland, the high temperature soared to 100°F or higher 20 times during that hot 1930 summer.

The Body Fights to Stay Cool

Heat waves can tax the human body to its limit. Ninety-degree-plus air temperatures place a huge burden on the body's effort to maintain its normal core temperature of 98.6°F. Fortunately, the human body has a built-in thermostat called the *hypothalamus,* located in the brain. When the hypothalamus senses a rise in blood temperature, it sends a signal to increase the circulation of blood to the skin. Bundles of tiny capillaries carry the blood to the skin's surface in order to release excess heat to the outside air.

As the heart works harder to pump more blood, sweat glands pour liquid onto the surface of the skin. Sweating cools the body as the water is removed by evaporation. (This is the same cooling sensation you feel as you step out of the shower or out of a pool and the body is exposed to drier air.)

Unfortunately, when there are high levels of moisture in the air (like on most summer days in Washington), sweat does little to cool the body because water is not evaporated from the skin. This is why meteorologists will often report the *heat index* during the summer. Devised by the National Weather Service, the heat index, sometimes called the *apparent temperature*, is a measure of how hot it really feels to the human body given the combination of temperature and humidity. For example, a temperature of 96°F and a humidity reading of 40 percent results in a heat index of 101°F. Increase the humidity reading to 50%, and it feels more like 108°F. If the humidity level reaches 60 percent, the heat index soars to 116°F – which is 20 degrees higher than the actual reading on the thermometer!

With prolonged exposure to heat indexes of 105°F or higher, heat-related illnesses may develop as the body continues to lose fluids through perspiration, while the body's internal temperature continues rising. Heat disorders range in severity from painful muscle spasms called *heat cramps* to life-threatening cases of *heat stroke*. Heat stroke represents a severe medical emergency and may occur when the body temperature reaches 106°F. Studies indicate that older people are at much higher risk for developing heat disorders. So, the heat may cause cramps in a 17-year old, heat exhaustion for someone in their 40's, and heat stroke for a person in their 60's.

The Bermuda-Azores High

Early mariners, including Christopher Columbus, referred to the vast expanse of the Atlantic Ocean around 30 degrees north latitude as the *horse latitudes*. Here the weather remained clear for weeks at a time, and there was often very little wind for sailing. Ships setting sail from Spain to the New World were often becalmed for weeks at these latitudes. Legend has it that Spanish sailing vessels transporting horses to the West Indies were forced to dump the animals overboard in order to conserve food and water – hence the name *horse latitudes*.

What explorers like Columbus didn't realize was that the region around 30 degrees north latitude was at the center of a sprawling area of high pressure called the *Bermuda-Azores High*. The Bermuda-Azores High is a semi-permanent area of high pressure in the North Atlantic Ocean that migrates east and west with the seasons. More commonly known as the Bermuda High, this subtropical ridge of high pressure plays a very significant role in summer weather for many areas east of the Continental Divide.

In the winter and early spring, high pressure is typically centered near the Azores Islands. With the approach of summer, the high strengthens and migrates westward. Usually centered near Bermuda in the summer, the massive high pressure system acts like a heat pump as its clockwise circulation taps hot, humid air from the Gulf of Mexico and the Atlantic Ocean and pumps it as far north as southern Canada. Under this regime, Washington and its suburbs will typically experience 85 to 90 degree temperatures, along with high humidity. This type of pattern usually leads to the familiar scattering of late afternoon and evening thunderstorms.

As the location and strength of the Bermuda High changes, so does our sultry summer weather. When it builds farther north and west, the Washington area is often situated beneath the dome of high pressure. As the air sinks beneath the high it is compressed as the higher atmospheric pressure near the surface squeezes the layer of air. The sinking air is warmed as it is compressed, leading to higher surface temperatures. Under these conditions, temperatures may soar to the century mark or even higher. Humidity levels, however, may

actually decrease somewhat as winds become very light and the moist flow of air from the Gulf of Mexico and Atlantic Ocean is cut off. Still, the nearly-calm winds result in very stagnant atmospheric conditions which trap pollutants like ozone and lead to very poor air quality.

These hot-weather patterns may persist for weeks, with only brief reprieves from the sweltering conditions. These reprieves usually occur when the ridge is suppressed a little farther south by an upper level trough moving through the Great Lakes toward New England. When this occurs, northerly surface winds may develop as a cold front pushes south of Washington, allowing cooler air to descend on the nation's capital. The result may be a two or three-day stretch of temperatures in the upper 70's or lower 80's, before the heat and humidity are cranked up again.

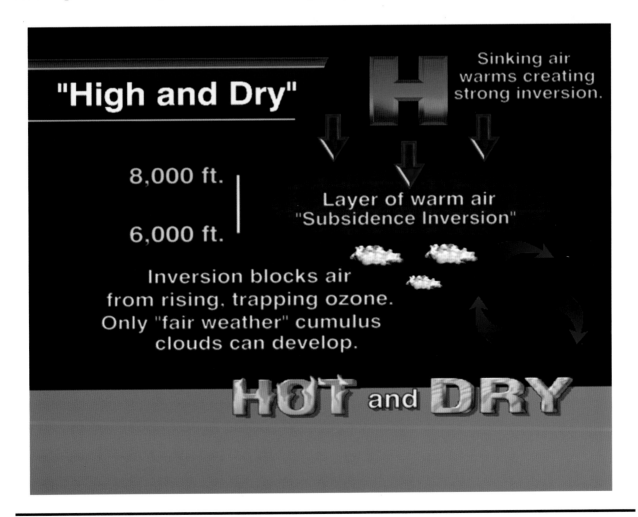

The Self-Sustaining Nature of Heat Waves

Heat waves, by nature, beget dry conditions, which in turn induce more heat. As air sinks under a massive area of high pressure, *subsidence inversions* are produced. This subsidence inversion refers to a very warm layer of air that forms in the middle portion of the atmosphere as air sinks from high above. The result is that the normal decrease of temperature with altitude is reversed. Consequently, as the sun bakes the Earth's surface and low-level air begins to rise, it loses its buoyancy once it encounters the warmer layer of air aloft. Sometimes the air rises high

dome of high pressure and continuing the cycle.

When subsidence inversions set up after a period of wet weather, they can trap low-level moisture, leading to stifling humidity levels. This was the case during the Chicago Heat Wave of 1995, when heat indexes soared to 125°F.

The Urban Heat Island Effect

One of the big differences between summer weather in urban and rural areas becomes evident when the sun goes down – there is no escaping the summer heat in the city. Highly urbanized areas, like Washington, D.C., are often warmer than the surrounding countryside, particularly at night. In fact, a close inspection of climatological data recorded at Reagan National Airport and at Dulles International Airport in Sterling, Virginia, located roughly 25 miles west-northwest of the city, illustrate this big disparity. High and low summer temperatures at Reagan National from June 1 to September 1 average 86.7°F and 69.3°F, respectively. In contrast, high and low summer temperatures at Dulles average 85.1°F and 62°F, respectively. While average high temperatures run a modest 2°F cooler at Dulles, average low temperatures run a striking 7°F cooler at Dulles. (Interestingly, winter temperatures show virtually the same spread.) While the Potomac River undoubtedly has a moderating influence on temperatures in Washington, there is another phenomenon that may explain these rather large disparities. It is called

Urban Heat Island
Overnight Temperature Profile

78
74
70
72
68

Downtown   Urban Residential   Park   Suburban Residential   Rural

enough for a few "fair weather" cumulus clouds to develop, but not high enough for clouds to continue to build into thunderstorms. As the ridge of high pressure continues to hold strong, the unabated sunshine dries out the soil, resulting in even less low-level moisture available to form clouds. Days on end of sunny skies and no rain further enhance the heat, thus reinforcing the

the *Urban Heat Island Effect.*

The Nation's Capital, like many big cities in America, resembles an asphalt forest. The streets, buildings, and monuments serve to enhance the heat. The hard, dark surfaces like tar, asphalt, concrete, and brick store much of the sun's energy during the day, and release that heat long after sunset. In addition, waste heat from buildings, cars, and trains is also continually released into the air. A study done by NASA scientists in Atlanta in May 1997 found that during the day, when the air temperature was 80°F downtown, the ground temperature was 120°F. They also measured rooftop temperatures of up to 170°F on some of Atlanta's buildings. This intense heat produced temperatures of up to 130°F in offices, rooms, and apartments that had no air conditioning. Although this particular study was conducted in Atlanta, it's reasonable to assume that a local study would yield similar results for Washington.

In contrast, rural areas are not as densely populated, are less developed, and feature bigger yards with more trees, shrubs, and grassy areas. As a result, solar energy is both absorbed and reflected by these green areas. The absorbed energy fosters evaporation of water from the vegetation and the soil. Because evaporation is a cooling process, it helps to offset the sun's heat during the day. At night, vegetated surfaces cool off much quicker than the city's concrete buildings. This allows nightly temperatures of rural areas to fall further than they do in highly urbanized areas.

Summer is No Time for Breathing Easy

Unfortunately, the lazy, hazy days of summer bring more than just heat and humidity to Washington. High temperatures, high humidity, bright sunshine, and light winds are the perfect recipe for a toxic brew of air pollutants – in particular, ozone. This ground-level ozone (or "bad ozone") should not be confused with stratospheric ozone (or "good ozone") found 10 to 15 miles above the Earth's surface. Stratospheric ozone filters out harmful UV radiation from the sun. Ground-level ozone is a product of hydrocarbons and nitrogen oxides – gases generated mainly by automobiles. In strong sunlight, these gases undergo chemical reactions to produce ozone.

A number of agencies, including the American Lung Association and the Environmental Protection Agency, believe ozone can cause irritation of the eyes, nose, and throat, and may also severely exacerbate symptoms related to bronchitis, heart

The water level in Maryland's Liberty Reservoir fell to 25 feet below normal levels during the Drought of 1999. Liberty reservoir supplies drinking water for much of north central Maryland. *WJLA*

disease, and asthma. Even healthy adults, who breathe in close to 3500 gallons of air in a single day, can have their lung functions reduced by up to 20 percent on days when high levels of ozone are present. Although ozone levels have declined in recent years in the Washington area, they still exceed federal standards on several days each summer.

Consequently, the Metropolitan Council of Governments issues ground-level ozone forecasts for the nation's capital and surrounding area, using a system of color codes: code green for good air quality; code yellow for moderate air quality; code orange for moderately unhealthy air quality; and finally, code red and purple for unhealthy levels of ozone. On code red, and especially code purple days, prolonged outdoor activity is not advised for anyone – particularly children and older adults with respiratory problems.

Coping with the Heat

The best way to cope with the simmer of summer is to stay indoors as much as possible. Plan any strenuous activities during the coolest part of the day, typically in the morning before 8:00 a.m. Wear lightweight, light-colored clothing, take frequent breaks out of the heat, and above all else, stay well hydrated. Regardless of whether you feel thirsty, drink plenty of water and avoid beverages with alcohol or caffeine. Also, be able to recognize the signs of heat exhaustion: heavy sweating; cool, moist, pale skin; headache; nausea; and dizziness. A person suffering from any one of these symptoms should get out of the sun immediately, cool the skin with a wet cloth or towel, and drink a half glass of cool water every 15 minutes. Listen to your body! It will tell you when you've had too much exposure to the sun.

Cooling off in the spray from a fire hydrant at 9th and Quincy Streets, NE, July 4, 1950. Washington's high temperature was 91°F. *Copyright Washington Post; Reprinted by permission of the D.C. Public Library*

HEAT WAVE EVENTS

The Record Hot Summer of 1930

The summer of 1930 made headlines due to unprecedented heat and drought that caused disastrous crop failures throughout the United States. The summer of 1930 ushered in the "Dust Bowl" era of unusually hot, dry summers that plagued the U.S. during much of the 1930's.

Washington area farmers were certainly not spared in 1930, as intense, prolonged hot spells gripped the region during late July and early August. The official temperature recorded on July 20 was 106°F, which holds the record as the highest temperature ever recorded in Washington. Unofficially, 110°F was recorded that same day on Pennsylvania Avenue and 108°F was recorded at the National Cathedral. The summer of 1930 also set the record for number of days where temperatures reached or exceeded 100°F at 11 days.

These high temperatures of over 100°F were recorded during two heat waves that occurred in late July and early August. The July heat wave high temperatures are as follows:

The surface weather map for the hottest day in Washington, July 20, 1930. A large high-pressure system was located over the southeast U.S. The all-time record high temperature in Washington of 106°F occurred that day. At the National Cathedral, the temperature reached 108°F. The low temperature was 76°F. *NOAA Library*

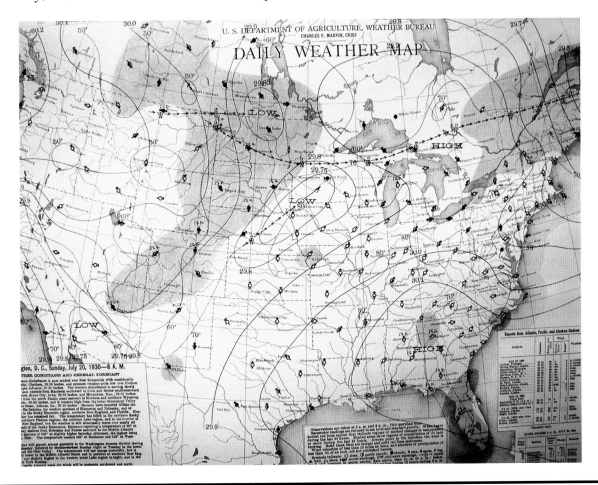

THE IRRITATING OVERTIME PARKER.

Heat wave cartoon that appeared in the Evening Star on July 28, 1930. The heat wave is pictured trying to break a "sitting record," emulating the popular flagpole sitters of the day. The summer of 1930 set the record for number of days that temperatures reached or exceeded 100°F at 11 days. The hottest temperature of 106°F occurred on July 20. Pulitzer Prize winner Clifford Berryman drew the cartoon. *Washingtoniana Division, D.C. Public Library*

July 19 - 102°F
July 20 - 106°F
July 21 - 103°F
July 22 - 100°F
July 23 - 94°F
July 24 - 93°F
July 25 - 100°F
July 26 - 100°F

The August heat wave high temperatures are as follows:

August 2 - 94°F
August 3 - 100°F
August 4 - 102°F
August 5 - 102°F
August 6 - 88°F
August 7 - 97°F
August 8 - 104°F
August 9 - 102°F

By the end of the summer, approximately 30 deaths in Washington were blamed on the heat and thousands more had died nationwide. Despite the discussions of global warming, there has never been another summer with heat waves that have equaled the heat wave of 1930.

The Evening Star's Heat Chaser hostess gives a motorcycle policeman a cold drink, August 4, 1936. Temperatures reached 95°F. The Heat Chasers provided cold drinks and refreshments during hot days in Washington. *Washingtoniana Division, D.C. Public Library*

Cooling off in the spray from a fire hydrant, June 24, 1948. Washington's high temperature was 98°F. *Copyright Washington Post; Reprinted by permission of the D.C. Public Library*

Cooling off in a fountain, June 21, 1949. Washington's high temperature was 95°F, with a dew point of 74°F. *Copyright Washington Post; Reprinted by permission of the D.C. Public Library*

Temperatures inside the armored Brinks' truck soar to over 110°F, July 3, 1950. The outside air temperature was a relatively cool 92°F. *Copyright Washington Post; Reprinted by permission of the D.C. Public Library*

Sunbathing at Pierce Mill on Rock Creek, July 1961. On ten of the last eleven days in July of 1961 the temperature reached 90°F or above. *Copyright Washington Post; Reprinted by permission of the D.C. Public Library*

A cold drink stand at Reservoir and MacArthur Blvd, September 3, 1961. Washington's high temperature hit 95°F. The temperatures for the first 6 days of the month exceeded 90°F. *Copyright Washington Post; Reprinted by permission of the D.C. Public Library*

Cooling off in the ice cream cooler, July 25, 1961. Washington's high temperature reached 93°F.

Copyright Washington Post; Reprinted by permission of the D.C. Public Library

Watching a Washington Senators' baseball game in D.C. Stadium, July 3, 1966. The game program turned into a hat to provide shade from the blazing sun. Washington's high temperature hit 101°F, and the low temperature was 75°F. *Copyright Washington Post; Reprinted by permission of the D.C. Public Library*

A George Washington University student wades in the fountain at 18ᵗʰ and E Streets, NW, June 28, 1969. Washington's high temperature soared to 100°F and the low temperature fell to only 76°F. *Copyright Washington Post; Reprinted by permission of the D.C. Public Library*

Drinking from a hose near Lafayette Park, July 6, 1977. Washington's high temperature reached 100°F, the low temperature fell to only 78°F, and the humidity was very high with a dew point of 74°F. *Copyright Washington Post; Reprinted by permission of the D.C. Public Library*

The Summer of 1980:
Prolonged Heat and Drought

The summer of 1980 featured a devastating, prolonged heat wave and drought across the central and eastern U.S. The heat wave was particularly intense in Texas, where Dallas recorded 100°F or higher on 69 days and had 100 consecutive days of 90°F or higher. The dry, stifling weather caused up to 5000 heat-related deaths, and wreaked $20 billion in damage to U.S. agriculture and related industries.

The sizzling temperatures set many Washington records, including the record for the hottest summer (June - August), with temperatures averaging 80°F. On every day from July 25 through August 14, the mercury hit 90°F or higher – an incredible 21 straight days! July 16 and 17 saw the highest readings, at 103°F and 102°F, respectively.

The Summer of 1988:
The Worst Heat and Drought since the 1930's

The summer of 1988 featured the worst drought in the U.S. Corn Belt region since the 1930's. A large, stationary ridge of high pressure was centered over the central U.S., which kept much of the nation hot and dry. At the height of the 1988 heat wave, 45% of the country was considered in serious drought conditions. Agricultural losses were estimated at a staggering $40 billion, with up to 10,000 deaths nationwide attributed to the heat. Corn and soybean prices almost doubled due to the tremendous crop losses.

In Washington, the heat wave began in early June and lasted through mid-August. Four days of 100°F+ temperatures occurred in July, with two more in August. And, similar to the summer

A hot and breezy day in Fairfax, Virginia, July 6, 1999. Washington's high temperature reached a sweltering 103°F.
Kevin Ambrose

of 1980, every day from July 29 to August 18 was 90°F or higher – 21 straight days.

Washington's hottest month: July 1993

The hottest month on record in Washington was July 1993. Washington's record heat coincided with devastating floods in the Midwest. July's average temperature in Washington was 83.1°F, with an average high temperature of 92.5°F and an average low temperature of 73.7°F. Four days during the month reached the 100°F mark, and 24 of the 31 days in the month reached 90°F or higher.

The Summer of 1999: Double Dose of Heat and Drought

The heat wave of 1999 gripped much of the eastern U.S. for weeks, breaking hundreds of long-standing daily and monthly temperature records and taxing the region's power grids. Rolling blackouts plagued the Northeast, particularly in Philadelphia, New York City, and Boston. The heat wave and drought went into record books as a multi-billion-dollar U.S. weather disaster, primarily due to massive agricultural losses. The human toll was costly, with an estimated 500 deaths attributed to the heat across the East.

The first taste of sweltering heat in Washington appeared in early June, when the mercury reached 98°F on three consecutive days from June 7 through June 9. The remainder of the month featured rather typical summer temperatures. But then, just in time for the July 4th fes-

tivities on the Washington Mall, the heat cranked up to a high temperature of 99°F. The searing heat continued virtually unabated through August 18, as temperatures at Reagan National Airport soared to 90°F or higher 38 times in July and August, and hit the elusive 100°F mark three times.

The extreme heat in Washington was also aggravated by severe drought conditions. During the hottest stretch of the summer (June 21-August 19), only 1.58 inches of rain fell in

Under the blazing sun in Washington, D.C., July 1999. The heat wave of 1999 gripped much of the eastern U.S. for weeks. *WJLA*

Washington. With conditions deteriorating and no relief in sight, Governor Parris Glendening issued mandatory water restrictions for the entire state on August 4. Meanwhile, much of Northern Virginia and the District of Columbia enacted voluntary water restrictions. Thanks in part to heavy rain from the remnants of Hurricane Floyd, most water restrictions were lifted in September.

Cooling off in a fire hydrant in Washington, D.C., summer of 1999. Washington has experienced many exceptionally hot summers during recent years. During the 70-year period from 1910 through 1979 there was only one summer (1930) that produced 15 or more 95°F+ days. Since 1980, there have been 9 such summers. *WJLA*

Kids beat the heat by playing in water fountains, August 8, 2001. The high temperature reached 97°F and the heat index soared to 105°F. The heat index is a measure of how hot it feels to the human body given the combination of temperature and humidity. *Larry Dowing/Reuters*

Launching a weather balloon from National Airport, April 23, 1942. The Weather Bureau staged this photograph to show that they actively employed women. *Copyright Washington Post; Reprinted by permission of the D.C. Public Library*

APPENDIX

Washington's Weather Means and Extremes

Temperatures

Average Temperatures:

Winter: **37.2°F**   Spring: **56.7°F**   Summer: **78.0°F**   Fall: **60.3°F**   Year: **55.1°F**

Record Temperatures:

Highest Temperature .... **106°F** ...... August 6, 1918 and July 20, 1930
Warmest Year .... **60.2°F** 1991
Lowest Temperature .... **–15.0°F** ...... February 11, 1899
Coolest Year .... **52.2°F** 1904
Longest Heat Wave .... **21 days** ...... July 29 – August 17, 1988
(90°F or higher)   July 25 – August 13, 1980

Rainfall

Average Rainfall (Includes melted snow/ice):

Winter: **8.55"**   Spring: **9.54"**   Summer: **11.09"**   Fall: **9.45"**   Year: **38.63"**

Record Rainfall:

Record 24-hour rain ...... **7.31"** .... August 11-12, 1928 (6:15 p.m. to 6:15 p.m.)
Wettest Month .... **17.45"** .... September 1935
Wettest Year .... **61.33"** 1889
Driest Month .... **Trace** .... October 1963
Driest Year .... **21.66"** 1930
Longest Dry Spell .... **33 days** .... August 7 – September 8, 1995

Snowfall

Average Annual Snowfall: 18.2"

Biggest Snowstorm .... **28.0"** .... January 27-29, 1922 (Knickerbocker Storm)
Snowiest Winter .... **54.4"** 1898-1899
Least Snowy Winter .... **0.10"** .... 1972-1973 and 1997-1998

High Winds/Severe Weather

Highest Sustained Wind ...... **78 mph** .... October 15, 1954   (Hurricane Hazel)
Highest Wind Gust .... **98 mph** .... October 15, 1954
Deadliest Tornado .... **17 killed** ...... November 9, 1926 (La Plata, Maryland)

BIBLIOGRAPHY

Ahrens, C.D. 1994. Meteorology Today: An Introduction to Weather, Climate, the Environment. 5th edition. Minneapolis/St. Paul: West Publishing.

Anthes, Richard A., Cahir, J.J., Fraser, A.B., Panofsky, H.A.. 1981. The Atmosphere. 3rd edition. Charles E. Merrill Publishing.

Asimov, Isaac. 1982. Asimov's Biographical Encyclopedia of Science and Technology, Second edition. Doubleday & Company, Inc.

Binczewski, George J. 1995, JOB, "The Point of a Monument: A History of the Aluminum Cap of the Washington Monument."

Brooks, H.E., C.A. Doswell III, J.E. Cooper. 1994. Weather Forecasting, 9, 606-618.

Catholic University, "The Chesapeake and Ohio Canal National Historic Park." Web Document.

Clean Air Partners. 2000-2001. "What is Ozone" and "Effects of Ozone." Web Documents.

Cobb, Hugh D. III. 1991. Weatherwise, August-September Issue: 24-29.

Dabney, Virginius. 1971. Virginia – The New Dominion, University Press of Virginia.

D'Aleo, Joe. Feb. 7, 2000. "Why Blocking Patterns Produce Weather Extremes," and "The Simmer of Summer." WSI Corporation Web Documents.

Davies-Jones, R., and H.E. Brooks. 1993. American Geophysical Union, Geophysical Monograph 79: 105-114.

Doe, Bruce R. 1999. American Geophysical Union, Vol. 80, No. 1: 1-5.

Doswell, C.A. III, H.E. Brooks, and R.A. Maddox. July 1995. "Flash Flood Forecasting: An Ingredients-Based Methodology." Weather and Forecasting.

Espinola, Mario 2001. "The Battle of Ox Hill". Espd.com Web Document.

Fassig, Oliver L. 1899. Maryland Weather Service, Volume I

Hedges, Robert, "The Encyclopedia of Dumfries, Virginia 1760-1771. Web Document.

Hickey, Donald R. 1995. The War of 1812. University of Illinois Press.

Jefferson, Thomas. 1954. Notes on the State of Virginia University of North Carolina Press.

Kaplan, John, and M. DeMaria. November 1995. Journal of Applied Meteorology: 2499-2512.

Kocin, Paul J., Uccellini, Louis W. 1990. Snowstorms Along the Northeastern Coast of the United States: 1955 to 1985. American Meteorological Society.

Laskin, David 1996. Braving the Elements: The Stormy History of American Weather. Anchor Books.

Lamb, H. 1995. Climate, History and the Modern World. Routledge.

Lott, Neal, 1993. "The Big One! A Review of the March 12-14, 1993 'Storm of the Century.'"

Ludlum, David M. 1982. The American Weather Book. Houghton Mifflin Company.

Ludlum, David M. Early American Winters I. American Meteorological Society.

Ludlum, David M. 1968. Early American Winters II. American Meteorological Society.

Mahon, John K. 1972. The War of 1812. University of Florida Press.

Mapp, Alf Johnson, 1985. The Virginia Experiment. Hamilton Press.

Maryland Weather Service, Volume I. 1899

Maryland Weather Service, Volume II 1906

Miller, Timothy L, 1999. "Urban Climatology and Air Quality: Heat Island". Web Document.

Moran, Joseph M., and M.D. Morgan. 1986. Meteorology: The Atmosphere and the Science of Weather. Burgess Publishing.

Morton, Richard Lee. 1960. Colonial Virginia. University of North Carolina Press.

NOAA. "National Weather Service Internet Weather Source: Heat Wave," Web Document.

National Weather Service, Louisville, KY. 2002. "Structure and Dynamics of Supercell Thunderstorms." Web document.

Peterson, Merrill D. 1986. The Portable Thomas Jefferson. Penguin Books.

Pitch, Anthony S. 1998. The Burning of Washington. Naval Institute Press.

Ray, Peter S. 1986. Mesoscale Meteorology and Forecasting. American Meteorological Society.

Readers Digest. 1989. Great Disasters. Readers Digest Association.

Reichelderfer, Francis W. 1946. Annual Meteorological Summary. U.S. Department of Commerce.

Rinehart, Ronald E. 1999. Radar for Meteorologists. 3rd Edition. Rinehart Publishing.

Roth, David; Cobb, Hugh. "Virginia Hurricane History," NOAA Web Document.

Shirley, John William, 1985. Sir Walter Ralegh. North Carolina Dept of Cultural Resources.

Smedlund, William S. 1994. Campfires of Georgia Troops, Kennesaw Mountain Press.

Smith, Bruce B. 2000. "What is a Waterspout?" National Weather Service Web Document.

Smith, John. 1982. The General History of Virginia. Southern Living Gallery.

The War of the Rebellion: A Compilation of the Official Records of the Union and Confederate Armies.

Trapp, R.J., and R. Davies-Jones. 1996. Journal of the American Meteorological Society, Preprints, 18th Conference on Severe Local Storms: 387-391.

Mergin, Bernard. 1996. Washington History, Slush Funds: 5-16.

Van Doren, Carl 1938. Benjamin Franklin.

Watson, Barbara McNaught. 1999-2001. "Washington Area Winters," "Storms of the Century in the Greater Washington-Baltimore Region," "Virginia Floods" and "September 24, 2001 Tornadoes." National Weather Service Web Documents.

Weisman, Morris L. 2001. Bulletin of the American Meteorological Society, Vol. 82, No. 1: 47-62; 97-114.

Welker, David 2002. Tempest at Ox Hill: The Battle of Chantilly, DaCapo Press.

Williams, Jack. 1997. USA Today Weather Book, Vintage Books.

Wood, W. J. 1990. Battles of Revolutionary War, Algonquin Books of Chapel Hill

Zynjuk, Linda D.; Majedi, Brenda Feit. January 1996 Floods Deliver Large Loads of Nutrients and Sediment to the Chesapeake Bay. U.S. Department of the Interior.

INDEX

A

Air Florida Plane Crash 86
Alberta Clipper 44
apparent temperature 226

B

bad ozone 229
Battle of Rappahannock 39
Bermuda High 175, 226
Bicentennial Winter 111
Blizzards
    1888 49
    1899 51
    1936 60
    1966 76
    1967 79
    1996 96
blocking high 115
Blocking Patterns 115

C

cA. *See also* Continental Arctic Air
Clayton, John 15
cold air damming 46
cold fronts 134
Cold Waves
    1780 31
    1912 118
    1917-1918 120
    1976-1977 120
    1980's and 1990's 126
condensation 176
Continental Arctic Air 112
Continental Polar Air 112
cP. *See also* Continental Polar Air
cyclogenesis 46

D

divergence 46
Doppler on Wheels 141
doppler radar 133
downdraft 135
dynamic pipe effect 138

E

El Niño 121
eye wall 177

F

flash flooding 200
Floods
    1889 203
    1924 206
    1936 206
    1942 208
    1969 211
    1972 216
    1985 218
    1996 218
    2001 221
    Potomac River 217
Franklin, Benjamin 15, 23
freezing rain 43
frostbite 117
Fujita Scale 140

G

Galilei, Galileo 17
Gates, Thomas 22
good ozone 229
Great Arctic Outbreak 111
ground-level ozone 229
gust front 136
Gustnadoes 139

H

hail 134
Hailstorms
    1975 153
    1999 158
heat cramps 226
heat index 225
heat island effect 126
heat stroke 226
Heat Waves
    1930 231
    1980 240
    1988 240
    1993 241
    1999 241
Historic Lightning Rods 148
hook echo 138
horse latitudes 226
Hurricanes
    1667 22
    1896 184
    1933 185
    Able 186
    Connie 191
    Hazel 187
    Hugo 178
hypothalamus 225
hypothermia 117

I

Indian Summer 18

J

James River 21
Jamestown Island 21
Jefferson, Thomas 16
Jet Stream 46

L

La Niña 121
landspouts 139
latent heat 176
Little Ice Age 18

M

Maritime Polar 114
meridional flow 115
mesocyclone 137
meteorological battle zone 43
mP. *See also* Maritime Polar
Mud March 37
Multi-cell storms 136

N

Nor'easters 44

O

orographic lift 134
outflow boundaries 198

ozone 229

P

Primrose 21
pulse storms 136

R

radiational cooling 112
rainshaft 164
Raleigh (Ralegh), Walter 20
rear-flank downdraft 138
Roanoke Island 20

S

Saffir Simpson Scale 177
shelf cloud 137
Siberian Express 113
single-cell storms 136
sleet 43
SLOSH 180
slush funds 107
Smith, John 14
Snow King 51
Snowstorms
  1772 25
  1776 27
  1857 34
  1909 (Taft Inaugural) 53
  1922 (Knickerbocker Snowstorm) 55
  1932 58
  1935 59
  1940 61
  1942 61
  1953 65
  1957 66
  1958 66, 69
  1961 (Kennedy Inaugural) 71
  1962 (Ash Wednesday) 73
  1971 80
  1979 81, 86
  1982 (Air Florida) 86
  1983 86
  1987 90, 92
  1993 93
  1994 95
  1999 101, 102
  2000 103

SPC 167
Spring tides 198
squall lines 136
stepped leader 134
Storm Prediction Center 167
storm surge 175
stratospheric ozone 229
streamer 134
subsidence inversions 228
Supercells 137
Surges 180

T

temperature inversion 43
The Great Fresh 25
thunderhead 134
Thunderstorms
  1862 (Battle of Ox Hill) 35
  1874 142
  1913 142
  1965 150
  1971 150
  1974 (TWA 514) 153
  1976 154
tidal bore 179
tidal flooding 198
tornado 131
Tornado Alley 132
Tornado Warning 141
Tornado Watch 141
Tornadoes
  1814 31
  1923 142
  1926 143
  1927 144
  1973 151
  2001 160
  2002 (La Plata) 165
training 197
tropical depression 177
Tropical Storms
  Agnes 193
  David 195
  Diane 191

U

updraft 135
urban flooding 198
Urban Heat Island Effect 228

urban snow hazard 107

V

Veterans Day Snowstorm 92

W

wall cloud 140
Ward, Robert 19
warm fronts 134
Washington, George 27
waterspout 139
Webster, Noah 19
White Hurricane 49
Williams, Samuel 19
Williamson, Hugh 19
Wind shear 137
wind-chill 117
wintry mix 43

Y

Year without a Summer 32

Z

zonal flow 115

ACKNOWLEDGMENTS

A special recognition to the following organizations for their assistance:

Thanks to WJLA for access to tape libraries and weather computers.

Thanks to The Washington Post for access to their photograph collections. A special thanks to Russell James.

Thanks to the NOAA Library in Silver Spring, Maryland for access to their collections of weather data, maps and photographs. A special thanks to Doria Grimes and Skip Theberge.

Thanks to the Library of Congress and the Martin Luther King Library for access to their photograph collections.

Thanks to the National Park Service for providing flood photographs and data of the Potomac River at Great Falls, Virginia. A special thanks to Barbara Perdew.

Thanks to the Associated Press for access to their photograph collection. A special thanks to Camille Ruggiero.

Thanks to Reuters for access to their photograph collection. A special thanks to Kathleen Bolger.

Thanks to Corbis for access to their photograph collection. A special thanks to Jemal Creary.

Many thanks to Weather Central, Inc for permission to use the Weather Central Genesis system to produce the high quality weather graphics.

A special thanks to the staff of the National Weather Service Forecast Office in Sterling, Virginia for their comments and suggestions. Their web site was also an invaluable source for local weather data, photographs and images.

Thanks to staff of the Storm Prediction Center in Norman, Oklahoma for data, comments, and suggestions.

Thanks to the Thomas Jefferson Foundation and the staff of Monticello. A special thanks to Jessica Tyree.

Many thanks to the following people for providing information for the book:

Barbara McNaught Watson, Warning Coordination Meteorologist, NWS Forecast Office, Sterling, VA
Barry Goldsmith, Meteorologist, NWS Forecast Office, Tampa, FL
Rich Thompson, Meteorologist/Lead Forecaster, Storm Prediction Center, Norman, OK
Will Shafer, Meteorologist, NOAA/Techniques Development Lab, Silver Spring, MD
Bernard Mergen, Professor of American Civilization, George Washington University
David Roth, Meteorologist HydroMet Prediction Center
Hugh Cobb, Meteorologist/Lead Forecaster, Tropical Analysis and Forecast Branch, Miami, FL
John Hart, Lead Forecaster, Storm Prediction Center, Norman, OK
Jeff Peters, Mesoscale/Outlook Forecaster, Storm Prediction Center, Norman, OK
Mario Espinola, Historian
Stan Rossen, ABC-7 Weather Watcher
Joseph Reintzel, ABC-7 Weather Watcher
Brian van de Graaff, ABC-7 Weather Producer

A note of appreciation to the following people for review of the accuracy of the text and/or editorial assistance:

H. Michael Mogil, Certified Consulting Meteorologist, "How the Weather Works"
Paul Knight, Instructor of Meteorology/Pennsylvania State Climatologist, Penn State University
John Hart, Lead Forecaster, Storm Prediction Center, Norman, OK
Doug Hill, ABC-7 Chief Meteorologist, Washington, D.C.
Jason Gough, KRIS-TV Meteorologist, Corpus Christi, Texas
Gerald Grossman, Meteorologist
Mario Espinola, Historian
Anne Ambrose, Editor
John Meyer, Editor
Doug Smith, Editor
Ovidio DeJesús, Layout

A very special thanks to the individual contributors of photographs:

James T. Bailey, Jr.; Eric Beach; Vincent K. Chan; Ovidio DeJesús; John DiCarlo, Ted L. Dutcher; James Foster; Kay Grahm; Marcia K. Hovenden; Katie Kahan; Steven Maciejewski; John Olexa, Jr; Barbara Perdew; Joseph Reintzel; Rick Schwartz; Gail Siegel; Michael Shore; Dr. Ming-Ying Wei; Steve Zubrick.

A snowy Capitol scene, February 10, 1926. The snowfall in Washington was 9.3 inches with a high temperature of 31°F and a low temperature of 25°F. *Library of Congress*